THE ART OF
URBAN
ASTRONOMY

Abigail Beall is a freelance science journalist. She studied Physics at Durham University before completing a master's in Science Journalism at City University in London. She grew up in Scotland and now lives in Leeds but spent most of 2018 travelling around Asia.

Find her on Twitter and Instagram @abbybeall

THE ART OF

URBAN
ASTRONOMY

A guide to stargazing
wherever you are

ABIGAIL BEALL

First published in Great Britain in 2019 by Trapeze
an imprint of The Orion Publishing Group Ltd
Carmelite House, 50 Victoria Embankment
London EC4Y 0DZ

An Hachette UK Company

1 3 5 7 9 10 8 6 4 2

A CIP catalogue record for this book is
available from the British Library.

ISBN (Hardback) 978 1 4091 9285 5
ISBN (eBook) 978 1 4091 9286 2

Designed and typeset by Goldust Design
Printed and bound in Great Britain by Clays Ltd, Elcograf S.p.A.

MIX
Paper from
responsible sources
FSC
www.fsc.org   FSC® C104740

www.orionbooks.co.uk

For Steff

Contents

INTRODUCTION

No matter where you are in the world, or how much money you have, everyone who can see has the ability to look up and view the night sky. You don't need to know a single thing about astronomy to feel a sense of wonder when you see a clear sky filled with stars. But when you do know, that feeling is much more powerful.

Living in a city, life can become stressful. With a busy job, dating life, trying to keep in touch with friends, maintain a social-media presence, go to the gym and remember to call your parents, sometimes it can feel like there is not enough time in the day for what you already do, let alone taking up a new hobby. But stargazing has been proven to reduce stress, improve mood and increase a feeling of awe. So take some time to look up at the sky, instead of looking down at your phone.

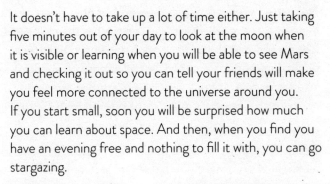

It doesn't have to take up a lot of time either. Just taking five minutes out of your day to look at the moon when it is visible or learning when you will be able to see Mars and checking it out so you can tell your friends will make you feel more connected to the universe around you. If you start small, soon you will be surprised how much you can learn about space. And then, when you find you have an evening free and nothing to fill it with, you can go stargazing.

Stargazing is not a solo hobby. You can take a date stargazing and impress them with your knowledge of the constellations, or get a group together, some blankets and a bottle of wine and go to see a meteor shower. Once you know the basics of stargazing in a city, the possibilities are endless.

This book is for anyone who has ever felt that sense of wonder when they look at the night sky, and would like to know a little more about what they are looking at.

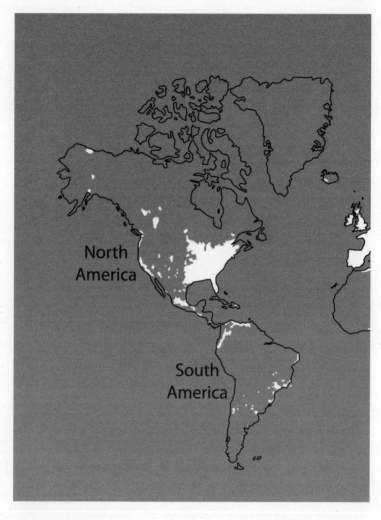

North
America

South
America

A map of light pollution

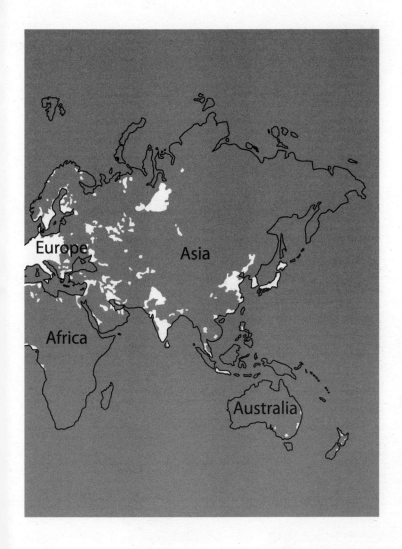

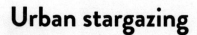

Urban stargazing

The chances are that you live in a city. Just over half of the world's population, 55 per cent of people, live in cities or urban areas, according to the UN, and this figure is only going to increase, to two thirds, in the next few decades. People in cities rarely take the time to look up, and many people will be surprised by how much there is to see when they do.

Cities are not the ideal place for stargazing but stargazing in a city isn't impossible. While faint stars just cannot compete with the thousands of homes, offices, restaurants and shops contributing to what lights up a city, there are many incredible things – meteors, the moon, planets and the brightest stars – that you can see from your own, light-polluted city. If you start to look up more and find it fascinating, you can start to explore more.

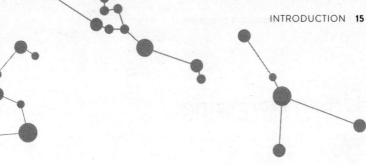

The first tip is to go as high as possible, away from street lights, or to make the sky as big as possible. If you can find your way to the roof of a tall building, that is a great place to start urban stargazing. If not, looking out of the highest window in your own building will do. Another good place is in the middle of a big park, especially one on the outskirts of the city.

Check what is going on in your local area. For example, amateur astronomy groups can be a great place to meet people as well as to learn about the night sky from someone who is familiar with it. Cities are often home to amazing resources where you can learn more about space, so make sure to go to local libraries, museums and observatories. Many observatories will offer talks or stargazing events, where you can go and learn about what is out there, and how to improve your stargazing skills.

You don't need a telescope, though: a pair of binoculars you can transport around is enough to see the details on the moon and even the moons around Jupiter, if the conditions are favourable.

Once you start studying the night sky in your city, it is something you can do anywhere you go. Whether you are going for a weekend away in the countryside or a holiday abroad, remember to look up. The techniques outlined in this book will work from any place in the world, so you have no excuse, and the darker skies the better, of course.

If you start to make a habit of checking when certain astronomical events will occur while you are away, you can add an extra element to your travels – looking at how the sky changes across the world. The first time I saw the moon from south of the equator was incredible, because it looks upside down – a different side is illuminated during the same part of the orbit. Close to the equator, the moon's crescent looks more like a boat, or a crescent at an angle.

Outside your own city, you may not have to venture far to find a dark sky. Resources online, listed in the Further Resources section at the end of this book, will tell you exactly where the darkest skies are in your area, so if you know there is an event coming up such as a meteor shower or an eclipse, it is worth travelling just that little distance away from light pollution because the amount you will be able to see will be worth it – as long as the weather behaves.

The speed of light

Every time you look at something, you are looking back in time. You can only see things because tiny particles of light, called photons, have travelled from that object and into your eyes. That process takes time, and in day-to-day life this hardly matters. The time it takes light to travel from your laptop or phone screen, or even from a bus you just missed a few metres away, or from the most distant object you can see on the horizon, is a fraction of a fraction of a second. Our brains cannot compute that delay, at all.

It remains true, however: for every single thing you look at, you are seeing a past version of that object. Mostly, it is a tiny fraction of time before you looked at it, which makes no difference, but not when it comes to space. Light moves faster than anything else in the universe, and through a vacuum, as space mostly is, photons cover a distance of 299,792,458 metres per second. This is incredibly fast, but when the distances involved get bigger and bigger, it starts to be noticeable. The speed of light means even light that reaches us from the moon has taken 1.3 seconds to get to us. From the sun, light takes eight minutes and twenty seconds, and from Neptune, it takes just over four hours.

Astronomers use the time light takes to travel in order to measure the vast distances in space. A 'light year' is a measurement for the distance light can cover in one year, which is 5.88 trillion miles or 9.46 trillion kilometres. Without using a measure like this, the scale of numbers involved would be ridiculous, and impossible to get your head around. That is what is so magical about looking into the night sky: everything you look at is a glimpse into the past. The sheer scale of the number of stars and the distances involved blows my mind, but it helps give you a sense of perspective too. Whenever you feel worried about something, just take a moment to think about how big the universe is.

The closest stars to us, other than the sun, are in a system of three stars called Alpha Centauri. These stars are just over four light years from the sun, which means the light we see when we look at those stars started its journey four years ago to reach our eyes. When you look into the sky, you are seeing the stars within our galaxy, which extends about 100,000 light years across. We don't lie in the middle of our galaxy, instead we sit on one of its spiral arms. There are at least 100 billion stars in our galaxy, the Milky Way, and with the naked eye we can only see a tiny fraction of this.

Many of the stars you see in the night sky might not even exist any more. They may have reached the end of their lives and run out of fuel, exploded or just fizzled out. But we can still see them twinkling in the sky, because we are looking into the past. With the naked eye, the most distant object you can find in the sky is the Andromeda Galaxy, our galactic neighbour, which is 2.5 million light years away. And with telescopes, astronomers can see back almost as far as the start of the universe, 13.8 billion light years away. If that isn't enough to give you a sense of perspective in the universe, I am not sure what could.

A brief history of astronomy

From the beginning of time

From the very earliest humans, we have always been fascinated by what the sky holds. The earliest civilisations thought the moon and the sun, and the way they rise and set, were controlled by the gods. It was also thought that the five planets visible in the sky had their own messages to tell as they moved around.

The Harappans, Maya and ancient Chinese used astronomy to keep track of time and to try to predict the future and this is where astrology began. Both the Babylonians and Egyptians used the phases of the moon to devise calendars and the former introduced a seven-day week, naming the days after the sun, moon and the five known planets.

From the eighth century BC, the Greeks were naming the constellations, documenting their positions and studying the way they move throughout the year. When Alexander the Great met Aristotle, in around 343 BC, he learned about astronomy. As he continued to travel and conquer different parts of the world, he spread what he had learned in Greece to those parts.

During the Dark Ages, much of the knowledge gathered in the Roman empire was lost, but Islamic astronomers in the Middle East preserved what they had learned from Greece and continued to further their knowledge and understanding.

Sixteenth century

In the sixteenth century, Nicolaus Copernicus was condemned a heretic when he announced that, from his observations, he believed the Earth moved around the sun and not the other way around. In 1539, Martin Luther, who led the Protestant Reformation in Germany, called Copernicus a 'fool' when he learned of the heliocentric theory. Copernicus's book was so controversial, it was not published until after his death.

Galileo Galilei was another pioneer of the heliocentric model of the solar system. He used a telescope, which is believed to have been first invented by Hans Lippershey in 1608, to study Jupiter and discover its four Galilean moons. This was the first discovery of something in the solar system that does not revolve around the sun. Obviously we know now that the Earth orbits around the sun, but when Galileo published a book that supported Copernicus's theory, he was put on trial for heresy by the Catholic Church. He was forced to read out an apology that took back his statement. He had been correct, of course, and years later the Vatican officially apologised.

Seventeenth century

In the early seventeenth century, Johannes Kepler developed a theory that correctly explains the way the planets move through the solar system. His theory was based on many of the observations by an astronomer called Tycho Brahe, who died in 1601. In 1572 and 1604 respectively, Tycho Brahe and Johannes Kepler saw what they believed to be the formation of a new star in the galaxy. Instead, what they were seeing was a supernova – the explosive death of a star resulting in it shining so brightly it could be seen from Earth for the first time.

In 1655, Dutch astronomer Christiaan Huygens spotted the rings around Saturn and its moon, Titan, for the first time. In 1664, Robert Hooke described the Great Red Spot on Jupiter.

Between 1665 and 1666, Isaac Newton sat under an apple tree on his own and came up with the idea of gravity, according to his biographer William Stukeley. For around twenty years, his theories of gravity existed only in his mind, until Edmond Halley visited him in Cambridge and asked questions about the orbit of the Earth around the sun. In 1682, Halley plotted the course of a comet through the sky. You might have heard of it.

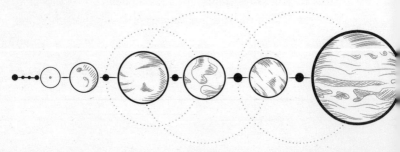

Eighteenth century

In 1726, Newton published the final edition of his life's work, *Principia Mathematica*. We still use Newton's laws of motion to describe how objects interact through the forces of gravity.

During this century, telescopes were developed further, especially by William Herschel, an English astronomer. He built around 400 telescopes during his lifetime, one of which he used to discover Uranus in 1781. In 1789, Herschel completed his largest telescope, which had a focal length of 12 metres (40 feet) and an aperture of 1.2 metres (4 feet). He used it to discover two of Saturn's moons, and pave the way for larger telescopes to be used to peer deeper into space.

Nineteenth century

After Herschel discovered Uranus, the first new planet to be found in hundreds of years, astronomers became interested in the solar system again. Hunting for planets between Mars and Jupiter began, and instead of planets they found small, faint objects like Ceres in 1801.

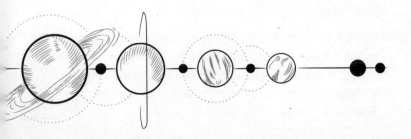

After Ceres was found, Pallas, Juno and Vesta were too: the first few known asteroids in the region we now call the asteroid belt. One astronomer was successful in the hunt for a new planet, however, and Neptune was found in 1846 by Johann Gottfried Galle with the help of calculations by Urbain Le Verrier and John Couch Adams.

Amateur and professional astronomy clubs were set up around the world, and members began to travel the world to watch events like eclipses. In 1874 and 1882, astronomers watched Venus transit in front of the sun.

Twentieth century

In the early twentieth century, work by Henrietta Leavitt helped reveal that some stars are much further away than others, even if they appear to have the same brightness in the sky. This was the start of the classification of stars.

In the 1920s, Edwin Hubble started to discover the distant galaxies, some of which were 100 million light years away. He also studied the movement of these galaxies and found the first experimental data to show that the universe was expanding.

In the 1930s, we began exploring the universe in other parts of the electromagnetic spectrum than just visible light: radio telescopes were invented.

In 1967, Jocelyn Bell Burnell discovered the first pulsar – a strange type of star that spins rapidly and shoots out energetic beams of radiation that can only be seen when the beam points towards us. For this reason, the stars look like they are pulsing, hence the name.

At the same time, the race was on between the USA and the Soviet Union to see who could make it to the moon. Fifty years ago, in 1969, humans set foot on the moon for the first time.

In 1995, the search for extra-terrestrial intelligence (SETI) began, listening to a specific part of the radio section in the electromagnetic spectrum where naturally occurring signals do not exist.

Present day

We have come a long way since the first humans looked up and wondered what was out there, but there is still a lot to learn about our universe. For example, we did not know of the existence of other galaxies a hundred years ago, so who knows what we will find within the next hundred years? In the 1920s, some astronomers still believed other galaxies were nebulae – clouds of dust that collect to form stars. Now, we believe there are at least 100 billion galaxies in the known universe.

Until 1991, we had not discovered any exoplanets – planets outside our own solar system. Now, as of February 2019, there are 3,976 confirmed exoplanets. Some of these might even have exomoons we could discover, and hints of the first exomoon discovery have been published already.

To this day, many discoveries are being made by citizen scientists, too. There is a vast array of websites that allow anyone to help contribute to astronomical research by classifying galaxies, for example, or finding particles that have landed on Earth from comets.

If you would like to become a name on the timeline of astronomy's history, a citizen-science project could be a good way. Some projects will include the participant's name on their scientific papers if they discover something new. In 2016, a pair of citizen scientists discovered a whole new galaxy, which has now been named after them. So what are you waiting for? Start looking up!

1

EVERYTHING EXCEPT STARS (THE CLOSER STUFF)

When people think of astronomy, they think of stargazing, but there is actually much more to the night sky than just stars. Everything that is not a star includes planets, meteors, the moon, aurorae, satellites – and in urban areas with light pollution, some of these can be the easiest things to spot. Some of the sky's most exciting spectacles happen incredibly close to home, in our very own atmosphere, so read on to find out more about what you can see in the night sky. We'll start with the closest to home, with aurorae and meteor showers, then move further away, to the moon and then the planets.

Aurorae

The aurorae are sometimes called the polar lights, northern lights or southern lights. Often green, sometimes red or purple, the spectacular lights are generated in the outer layers of Earth's atmosphere. In the northern hemisphere, their official name is aurora borealis, while in the southern hemisphere it is aurora australis. Whatever you call them, seeing the shows of differently coloured lights dancing across the sky is an experience many people travel hundreds of miles to get the chance to see.

Historically, the moving-light displays have been linked to angry gods or the blood of martyrs flowing up into the sky. Rock paintings in France show drawings of what appear to be the northern lights, dating back 30,000 years. In ancient China, they were linked to animals like dragons, and in Norway they were thought to be a bridge between the gods and the Earth.

To this day, these spectacular shows of light provide unforgettable experiences for those who are lucky enough to witness them. While we know exactly what causes them now, they remain magical. Much like anything in astronomy, the experience of witnessing aurorae is even more special if you understand what is causing them in the first place. In this case, it is particles thrown our way by the sun.

**BEST PLACES TO SEE THEM
IN THE WORLD**

Europe: Iceland, the north of
Norway, Sweden and Finland

North America: Yukon and Alaska

South America: Patagonia,
Falkland Islands

Oceania: Queenstown,
Tasmania and Victoria.

What are they?

Deep in the Earth's core, there is a ball of molten iron
that swirls around as our planet rotates and generates
a current, which produces a magnetic field around the
Earth. The magnetic-field lines roughly loop round the
Earth from the south to the north magnetic pole. This
very same magnetic field is what moves the needle on
your compass to point north. But the field also protects
us from high-energy cosmic rays and charged particles
flying through space.

The sun is amazing, as it provides us with light and heat:
without it, we would just not exist. But not everything
that comes from its surface is good. Charged particles,

like electrons and protons, shoot out at times when the sun is particularly active, from the corona, the outer layer of the sun. This happens during something called a coronal mass ejection. The particles move at high speeds, covering a distance of around 400 km every second, which means they take around forty hours, or just under two days, to reach Earth from the sun. Because these particles have a charge, they change direction when they hit the Earth's magnetic field. Mostly they are deflected away from Earth, but some particles enter the atmosphere at the point where the Earth's magnetic field is weakest – the north and south poles.

When the protons and electrons enter the upper atmosphere, they collide with gas particles in the air. This collision provides the gas particles with extra energy and excites electrons within them to a higher energy state. This higher energy state is unstable, so the electron will fall back down and release the energy in a photon, or a light particle. This is the same process that creates light in neon bulbs and street lamps.

The gaps between the energy levels in each atom are different. This means it takes a different amount of energy to excite an electron in an oxygen atom, for example, compared to a nitrogen atom. The result is that when these two different atoms release photons they have different energies, so we see this as different colours of light.

The height at which the charged particles from the sun hit the gas particles in the atmosphere also changes the colour we see. The most common colour of aurorae is pale yellow or green, and this comes from oxygen particles around 60 miles (97 km) above the ground; the rarer red aurorae come from oxygen higher up in the atmosphere, around 200 miles (322 km) above ground, and purple or blue aurorae come from nitrogen.

How to see them

Most people will have to travel if they want to witness these stunning light shows. If you live in the UK, especially in Scotland or the north of England, you could be lucky. Similarly, if you live in the north of America you may see them occasionally.

Wherever you are, there are a few steps you can take to increase your chances of seeing the aurorae. Firstly, check the weather and keep an eye on when you might have a cloud-free night sky, to maximise your chances of seeing them. As you know, aurorae are created by particles from the sun, so knowing when the sun's activity is higher helps predict when you might see them. Much like weather on Earth, solar activity is hard to predict. However, there are a huge array of websites and resources online that

BEST TIME TO SPOT AURORAE

In the winter, when the nights are darkest, but this often means it is likely to be raining or even snowing. A clear winter's night is the perfect time to see the aurorae.

will help you monitor what the solar activity is likely to be like in the coming day, or even week. The USA's National Oceanic and Atmospheric Administration (NOAA) website provides thirty-minute predictions along with the sun's activity for the past three days, which might give an indication of the coming activity. This is measured in a number called the planetary K index, or K-p, and is on a scale from zero to nine. The greater the K-p value, the higher the sun's activity.

If it's a clear, dark night and the sun looks like it will be active, get yourself to a dark spot. This might mean having to drive away from the city or village you are staying in, to get away from light pollution. Then you need to wait and let your eyes adjust. No doubt you will have seen countless photographs of the aurorae in stunning, bright colours. This is not what you see with your eyes. In reality, it is a much subtler phenomenon, and can even be quite tricky to see, the first time you try. For this reason, if you are in a touristy area popular for aurorae, consider going on a tour with a trained guide who will be able to show you exactly what you are looking for and who will help you capture those all-important photographs, too.

Certain phone applications can help you spot the aurorae by taking a long-exposure photograph of the sky and bringing out certain colours. This can help you find them, but make sure you take time to appreciate them without looking through the lens. Remember, they move quickly so you don't want to get too distracted getting the perfect photograph. Instead, just sit back and marvel at the incredible show of lights put on for us by our sun.

Meteor showers

Although they are often called shooting stars, a meteor could not be more different from a star if it tried. Stars, like the sun, are some of the biggest and brightest objects in the universe, while shooting stars, or meteors, on the other hand, are created by something tiny, usually between the size of a grain of sand and a grain of rice. When they enter the Earth's atmosphere and bump into air particles, they experience a huge amount of friction. This slows them down and makes them burn brightly. This flash is what we call a shooting star, or a meteor.

Some meteors are caused by dust or rock from asteroids and some from comets. Asteroids and comets are all around 4.5 billion years old, roughly the same age as the solar system. When the solar system formed, bits of dust, rock and ice started to clump together and orbit the sun. Eventually, these became the planets. However, during this process there were a lot of collisions and loads of bits of rock and dust were left over; these are asteroids and comets.

Asteroids and comets, like the planets, orbit the sun, but their orbits can be strange and unstable because of how low their masses are. Their size can vary from the size of a pebble to almost 1,000 km in diameter. They can sometimes be caught off guard and shoot straight towards the sun, passing Earth in the process.

The difference between asteroids and comets is what they are made of; asteroids are made of rock and metal, because they formed closer to the sun. Most asteroids orbit the sun between Mars, the furthest rocky planet, and Jupiter, the closest gas giant, in what is called the asteroid belt. Ceres, which is technically an asteroid but also sometimes called a dwarf planet, is 945 km in diameter, making it the biggest object in the asteroid belt.

As comets started out life further away from the sun, they are colder, which means they are made up of ice and rock. If a comet's orbit brings it near the sun, its ice turns from solid into gas, or vaporises, and this creates a tail. They can also be surrounded by a cloud of vaporised ice called a coma.

When asteroids or comets speed towards the sun, they leave behind a trail of dust and these trails stay around. These bits of rock, that can be up to 0.6 miles (1 km) in diameter, are called meteoroids, so when Earth moves through these trails of dust, we know to expect a meteor shower. If any meteoroids are large enough to survive entry into the Earth's atmosphere without being destroyed completely, the bits of rock that hit the ground are called meteorites.

Earth moves through the aftermath of certain comets at specific times of year, so spotting meteors can be easy if you know when to look.

How to watch a meteor shower

Meteor showers tend to be named after the constellation in which they appear to start in the sky. This spot is called the radiant.

For example, if you want to watch the Geminids meteor shower, find the Gemini constellation. If you want to watch the Perseids meteor shower, look at Perseus. You can probably guess where the Orionids, which came from Halley's comet, start from in the sky.

In January, there are the Quadrantids, followed by the Lyrids in April. There are two Aquarids meteor showers, in May and July. In August, there are the Perseids, followed by the Piscids in September, the Orionids in October, the Taurids and Leonids in November, and the Geminids and Ursids in December.

There are many online resources, listed in the Further Resources section of this book, that will tell you when exactly to expect to witness specific meteor showers, even down to the time of day to see them. It is normally best after midnight.

If you are interested, at the start of each year mark the date in your diary, and nearer the time check weather conditions and the time it will pass, to know exactly when to expect to see the meteors. It is also important to note what the moon is up to then. If there is a full moon, it will be bright in the sky which will make it harder to spot the meteors. The best time to see a meteor shower is during new moon.

To see a meteor shower, it is best to try and get as far away from light pollution as possible. Go to the middle of an open park, or up a hill if you can. If not, try to get to the highest point in your building, or a nearby rooftop. Find the constellation you are looking for, with the help of star charts, or an app if you feel like it, and get comfortable. If you are unsure whether the area you have chosen is dark enough, look for the Little Bear constellation, also known as Ursa Minor (chapter 3 in this book explains how to find it): as a general rule, if you can see this constellation, you will be able to see the meteors, if they are there.

Then it's time to lie back and wait for the show to begin. Maybe bring some friends, a hot water bottle or a bottle of wine. If it is your first time watching a meteor shower, do not try to take photographs as you will only struggle to get the brief flashes on camera, and you won't enjoy the meteor shower. Until you have seen enough meteor showers that you don't mind one being ruined by the frustration of not capturing it on camera, there is no point trying take photographs.

Although we know when to look for meteor showers, random events do occur, too. These are called sporadics. If you are stargazing on any given night when a meteor shower is not expected, a general rule is that you can expect to see between five and ten sporadic meteors. They can be seen at any point in the sky, travelling in any direction.

Larger meteors can technically strike at any time. For example, in 2013 a near-Earth asteroid hit the Russian town of Chelyabinsk and lit up the sky as it entered the atmosphere. It is thought to have entered Earth's atmosphere with a diameter of almost 20 metres, but the largest meteorite found after the impact measured 60 cm across. Nobody died from the impact, but hundreds of people were injured; mainly because of windows breaking due to the explosion, and eye damage from looking at the meteor which shone thirty times brighter than the sun.

Comets

Meteor showers happen when Earth passes through the tail of comets or asteroids, but it is sometimes possible to witness comets passing Earth, too. Sometimes these can be unexpected, but sometimes we know they are coming. For example, the famous Halley's Comet which passes Earth roughly every seventy-five years is expected to come by again in 2061. Better put that in the diary now, while you remember.

Satellites

When you think of space, you might imagine the Earth as it is shown in most diagrams – surrounded by its atmosphere and then, after that, nothing. Unfortunately, this is not the case. There are thousands of man-made objects orbiting the Earth. Some of these are working satellites and some are just old pieces of satellite that stopped working years ago. Put simply, Earth is cluttered with space junk.

Man-made objects like satellites, aircraft and weather balloons can be seen after the sun sets, and some may appear similar to a meteor. So how can you tell whether you've spotted a meteor or a satellite? Firstly, aircraft are easy to identify – red and green lights on each side and a blinking white light are a giveaway. If the plane is moving straight towards you, it may appear like a star getting brighter, so perhaps try not to stargaze near an airport.

Satellites tend only to be visible just after sunset or just before sunrise, because they do not have a light source of their own but only reflect light from the sun. Satellites tend to move slowly but steadily across the sky, much faster than stars or planets. Unlike meteors, which appear as a single flash of moving light, most satellites normally maintain a constant brightness as they move across the sky. Some may appear to increase in brightness or even flash if they are spinning around, but this will result in a series of flashes, unlike for a meteor.

One satellite that can be fun to spot is the International Space Station (ISS), which is among the brightest of the satellites because of its size. The ISS has been constantly occupied since 2000, but the first stage of the satellite was launched into orbit in 1998. It measures 108 metres, or 357 feet, end to end. It orbits at a height of 253 miles (408 km) above the Earth's surface.

The ISS takes ninety-two minutes to orbit the Earth, and completes fifteen and a half orbits every day, but it is not constantly circling around the same route.

To spot the space station, it has to be dark, because it is not bright enough to see during the day. The chance of this happening where you are varies from once a month to a few times in one week. NASA has a great website called 'Spot the Station' which will tell you exactly when you can expect to see the ISS overhead where you are.

Moon

The moon is almost as old as Earth itself. It's thought the moon was created around 4.5 billion years ago, when a large rock the size of Mars crashed into our planet. The debris left over started to clump together under gravity and began orbiting Earth. This is when the moon was born.

Moon's orbit

The moon orbits Earth in a synchronous orbit. This means it takes the same amount of time to complete a full rotation as it does to orbit around the Earth. The result is that we always see the same side of the moon when we look up and this is what we call the near side of the moon. The far side of the moon is often mistakenly referred to as the dark side, but this is not the case as it gets just as much sun as the near side, it's just we can never see it.

The orbit of the moon around Earth is not a perfect circle, it is elliptical. It's also constantly changing because of tugs of gravity from the sun and Earth, so one month it may be closer to Earth than the next. The point during the moon's orbit when it is closest to Earth is called perigee, and the furthest-away point in the orbit is the apogee.

The moon's distance from Earth varies during its orbit from roughly 250,000 miles away to 220,000 miles away. But this changes, too, as from one month to the next the moon can be 300 miles closer to Earth during its closest point. On average, the distance from the centre of the Earth to the centre of the moon is 238,000 miles. This change makes a large difference in how the moon appears to us. A full moon during perigee will appear around 30 per cent brighter and 14 per cent larger than an average full moon. When a full moon occurs during perigee it is called a supermoon.

Historically, the moon has been incredibly important to humans. Before electricity was discovered, moonlight allowed farmers to get up earlier and plough their fields by the light of the moon. But it has not always been a good thing. The term 'lunatic' was first coined because people were thought to have diseases relating to the moon. People were described as lunatics in books including the Bible when they fell to the ground shaking during the full moon. It is now thought this could have been a form of epilepsy, triggered by the bright light of a full moon.

As recently as the past couple of hundred years, people in mental institutions believed the moon was linked to poor mental health. Stories of patients becoming more agitated and restless during a full moon were common. However, it is now believed that their sleep was simply disturbed by the bright light of the full moon. These buildings were far from any others and so were in deep darkness. The patients' rooms did not have curtains, as this was a suicide hazard, so the light of a full moon could have easily disturbed at least one patient's sleep. Now, with the amount of light pollution and artificial light, it's unlikely our mental health would be as affected by the moon. But stargazing definitely can be.

Because of the historical ties to the moon, the full moon is given different names depending on the month it can be seen in. There are hundreds of different names for the full moon in each month, depending on where they originated. Today, there are often reports in the media when a supermoon is expected to be seen, referring to the moon by its Native American name.

Full-moon names

In January, Native Americans named the full moon the 'wolf moon'. This is said to be because the wolves were particularly hungry in January, after months of winter without food, so they howled the loudest during that time. It's thought to also have been called the 'old moon' or 'ice moon'.

In February, the cold weather in north America gave the full moon its title of 'snow moon'. Other names for February's full moon include 'storm moon' or 'hunger moon'.

In March, the icy ground started to thaw and worm trails appeared, which is why March's full moon is the 'worm moon'. In April, the blossom gave the moon its nickname of 'pink moon'.

In May, flowers came for the 'flower moon'. In June, the harvest gave the moon its name 'strawberry moon'. July is the 'buck moon' because the deer begin to regrow their antlers, and August is the 'sturgeon moon' because the fish were more plentiful during that month.

In September, the full moon is the 'harvest moon' because the light meant farmers could harvest their crops by its light. In October, it's the 'hunter's moon' because the newly harvested farms were the perfect arena to hunt foxes and deer. In November, it's the 'beaver moon' and December is the 'cold moon'.

As the moon's orbit is not exactly in line with our calendar months, there will sometimes be a month during which there are two full moons. The second of these is called the 'blue moon'.

Phases of the moon

The moon takes 27.3 days to complete one orbit of the Earth. During this time, it cycles from a new moon, where nothing can be seen, to a full moon, where you can see the whole thing, to a new moon again. This is because the moon is not a source of light itself, so only reflects the light from the sun. When the moon is between the Earth and the sun, no light is reflected. This is new moon. When Earth is between the sun and the moon, we can see a full circle, or full moon.

The phases of the moon change between new moon, waxing crescent, first quarter, waxing gibbous, full moon, waning gibbous, last quarter, waning crescent and new moon again. The phases may seem complicated but there are a few rules to remember:

- Between new moon and full moon, when the moon is getting more illuminated, it is waxing.

- After full moon and before new moon, when it is becoming less illuminated, it is waning.

- Confusingly, a quarter is when we can see half a circle in the sky – but remember, this is a quarter of the moon's surface because there is a half we can never see.

- A crescent is a small slither and a gibbous is almost a full moon with a small slither taken away.

New moon is often considered the best time of the month to go stargazing because the moon is incredibly bright when it is in the sky, and this light can overshadow other astronomical events. For example, meteor showers are difficult to see during a full moon, as are aurorae.

During a full moon, you can look and see evidence of our ever-changing solar system: impact craters. Compared to Earth, the moon hardly has an atmosphere. A mixture of gases has been detected surrounding the moon's surface, but it is tiny compared to what surrounds Earth. At sea level on Earth, each cubic centimetre of space contains 10,000,000,000,000,000,000 air molecules. The moon's atmosphere has less than 1,000,000 molecules in the same volume – this is what we would consider a vacuum on Earth.

THE BEST THINGS TO SEE ON OR NEAR A NEW MOON

A new moon means there is no light coming from the moon, making the skies darker. This means it's easier to see fainter constellations you might not otherwise see, and phenomena like the northern lights and meteor showers. It can be a good time to look for aurorae, or deep-sky objects, like the Andromeda Galaxy, too.

**THE BEST
THINGS TO SEE ON OR
NEAR A FULL MOON**

A full moon means there is a bright light in the sky, so only the brightest constellations will be visible. But grab a pair of binoculars and you can study the craters of the moon up close. You can also easily see the planets that are in the sky at that time, and other bright objects like the ISS if it happens to be passing.

That means when bits of rock and dust from asteroids and comets fly towards the moon, there is nothing to slow them down or burn them up. The moon is not geologically active, either, unlike Earth. This means the moon's surface is covered with evidence of billions of years of these impacts happening. These craters can be seen with the naked eye, and the best time to find them is two days before or two days after the full moon, because the light from the sun highlights them best on those days.

Start with Copernicus, which looks like a bright white dot but is in fact a 58-mile-wide (93 km) crater near the middle of the moon as we can see it from Earth. It is the easiest to spot with the naked eye. Next to Copernicus is a smaller crater called Kepler, but this can just appear as a fuzzy spot unless you have binoculars. The two craters are surrounded by darker patches, making them easy to identify. Another crater to find is Aristarchus, which is 25 miles (40 km) across and found just to the left of Copernicus. Then have a look to the south of the moon to find the Tycho crater.

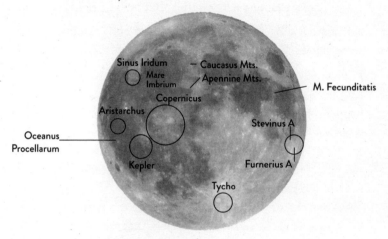

So next time the bright full moon prevents you from witnessing a meteor shower, look at those craters and remember that if you were actually on the moon, you would never see a meteor shower. Every time a meteor shower happens on Earth, bits of rocks and dust shower down on to the surface of the moon.

However, the moon can provide exciting opportunities to observe when its orbit brings it into the shadow created by Earth, or when it blocks the light from the sun. These are called eclipses.

Eclipses

As mentioned, a full moon is when the moon is at the opposite side of Earth from the sun, so its full surface is illuminated, and a new moon is when the moon is on the same side as the sun, so there is no reflection.

A lunar eclipse occurs when the Earth's shadow blocks light from the sun from reaching the moon. This can only happen when there is a full moon, because the Earth is between the moon and the sun. So why is there not a lunar eclipse every month? The plane of orbit of the moon around the Earth is not the same as the plane of orbit of the Earth around the sun. Imagine you draw the Earth and the sun on a piece of paper, with the Earth orbiting around the sun in a flat circle – that means the paper is the plane of the sun's orbit. If you wanted to draw the moon's orbit, you could not draw it on the same piece of paper. You'd need to get a new piece of paper and draw the moon in an orbit around the Earth, and intersect the two pieces of paper at an angle.

Sometimes, the moon will pass into the plane of the Earth's orbit around the Sun, meaning the sun, Earth and moon line up in a straight line. This is called syzygy and eclipses only occur when syzygy happens. When the sun, Earth and moon line up during full moon, a lunar eclipse happens. As the moon passes into Earth's shadow, it lights up in a reddish glow and this is called a blood moon.

During a new moon in syzygy, when the moon passes in front of the sun and blocks its light from reaching Earth, a solar eclipse happens. Because the moon is so small compared to the sun, the moon's shadow does not hit all of Earth at once – so total eclipses can only be seen from certain parts of the world; for example, a solar eclipse in July 2019 was visible from some parts of Chile and Argentina. When the darkest part of the moon's shadow hits Earth, there is a total solar eclipse. However, often the moon only blocks part of the sun's light and this is a partial solar eclipse.

Luckily, we understand the orbits of the moon and Earth well enough to predict lunar and solar eclipses, and they can be exciting events to watch. Wherever you are in the world, there are many online resources listed at the end of the book that will tell you when you are next likely to see a total solar eclipse.

HOW TO SPOT THE PLANETS

The planets move around, so there is no easy rule for finding them. Use websites listed in the Further Resources section to determine which planets will be visible, and where they will be, at any given time.

MERCURY
Look just after sunset in the west, or just before sunrise in the east

VENUS
This bright planet is often mistaken for a UFO, and is usually near the sun in the sky

MARS
Has a reddish glow and can be found anywhere along the ecliptic

JUPITER
Glows blue/white and appears as the third brightest 'star' in the sky

SATURN
Appears pale yellow and can often be mistaken for a star

URANUS
Cannot be seen with the naked eye

NEPTUNE
Cannot be seen with the naked eye

Planets

Each of the planets has its own agenda – each is orbiting the sun on its own path, and rotating at the same time, just like Earth. What that means is, unlike the stars, the planets move through the sky. When you can see more than one, they form a straight line in the sky, travelling along in the same direction as the moon and the sun – this means if you know where to look, you can clearly spot the planets. Even through the glow of a light-polluted city, the lights reflected from Venus, Mars and Jupiter shine brightly in the sky.

If you look up at the night sky regularly, you will notice a few particularly bright stars that seem to move through the night, faster than the rest of the stars. The likelihood is that these are the planets, and when they are visible in the night sky, they can be some of the brightest objects you will see. Spotting the planets is an exciting way to get started with stargazing because there are telltale ways to pick them out. Once you know, you can impress all your friends.

The ancient Greeks called the planets wandering stars, or *planetes asteres*, which is where the name 'planet' comes from. This is because their orbits mean they move independently of the stars in the night sky, which only appear to be moving because of the Earth's orbit and rotation. The Earth orbits the sun in a plane. From Earth, this looks like the sun traces a path across the sky during the day, rising in the east and setting in the west. This line the sun traces out is called the ecliptic. Because the planets roughly orbit in the same plane as Earth, when there is more than one planet visible to the naked eye, they roughly follow the path of the sun in the sky.

Finding the planets can be tricky at first because the position of each planet changes relative to the stars each day, as they are also orbiting the sun at their own speeds. Because of their movements, you need to be up to date about the planets' positions before you look for them. Any stargazing app will tell you which planets are visible in the sky and where to find them on any given day.

Just after the sun formed, for about 100 million years it was surrounded by a cloud of dust and gas. The gas evaporated and the dust began to clump together, and this is what makes up the planets in the solar system today. Exactly how the planets were formed is not fully understood, but at the time of writing, astronomers have three major models for how the planets were formed. While the most popular model explains the inner planets, the rocky planets, it does not cater to the gas giants so well.

What we can be sure of is that the solar system is 4.5 billion years old. There are eight planets: Mercury, Venus, Earth, Mars, Jupiter, Saturn, Uranus and Neptune. Controversially, Pluto was officially declassified as a planet in 2006 to become a dwarf planet, of which there are five officially classified but many more are expected to be discovered.

The planets you can see with the naked eye are the rocky Mercury, Venus and Mars, and the gas giants Jupiter and Saturn. Mercury and Venus, because they are closer to the Sun than we are, are usually found next to the sun in the sky, meaning they are best seen just after sunset or before sunrise. The rest of the planets are not necessarily close to the sun in the sky; they can often be 180 degrees away, for example. The ice giants, Uranus and Neptune, cannot be seen with the naked eye at all, but they can be found with a telescope.

Mercury

Because it is usually close to the sun in the sky, Mercury normally sets in the west around an hour after the sun, so dusk is a good time to look for it. In the morning, it rises in the east about an hour before sunrise, so you can find it then, too. It's the smallest planet we can see with the naked eye, and it twinkles like a star, with a slightly yellow tone.

Mercury is the closest planet to the sun, orbiting in an elliptical shape with an average distance of 36 million miles (58 million km), which means it takes light just over three minutes to reach Mercury from the sun. The planet takes eighty-eight Earth days to complete its orbit around the sun, and spins very slowly. For every two orbits around sun, Mercury spins three times.

Mercury is a rocky planet, like Earth, and although it's the closest to the sun it is not the hottest planet. Without much of an atmosphere to trap the heat, Mercury's temperatures vary drastically from a daytime 427 degrees Celsius to a nighttime −173 degrees Celsius. Mercury is the smallest planet in the solar system, and its surface is the most covered in impact craters.

Sometimes, Mercury can be spotted transiting in front of the sun, but this is a rare event which happens thirteen times in every century. At the time of publication, the next time this will happen is 11 November 2019.

Venus

Our closest planetary neighbour looks big and silver in the night sky. When it's visible, it becomes the second brightest object in the sky, after the moon. Some say it has been mistaken for UFOs because it is so large, silver and bright.

Even though Venus is further from the sun than Mercury, its surface is hotter because of its thick atmosphere filled with greenhouse gases. The effect of these gases is that the surface of Venus has heated up to an average of 464 degrees Celsius. Light takes six minutes to reach Venus from the sun.

Venus is somewhat the odd one out of the solar system, too, as it spins in the opposite direction to most of the other planets – east to west. It also spins incredibly slowly. It takes Venus 243 Earth days to rotate around on its axis, and it takes 225 days to complete an orbit of the sun. This means a year on Venus is shorter than a day on Venus.

Mars

Known as the red planet, Mars lives up to its name in the night sky, as it glows with a reddish hue to the naked eye. Depending on how close it is to Earth, Mars can become brighter than Jupiter in the night sky; this happened in 2003 and 2018. For most of 2019, Mars is relatively far from Earth, but it moves closer roughly every two years. Light takes almost thirteen minutes to reach Mars from the sun.

Mars is quite small – just over half the mass of Earth – but it is an exciting planet. It is thought that billions of years ago, Mars was covered in oceans and rivers of liquid water, but today there is no liquid water on its surface. In 2018, a NASA rover found organic molecules that could mean there was once life on the planet. In the same year, what is thought to be a lake of water was found hiding one mile (1.6 km) underneath its surface at its south pole, using radar technology, by the European Space Agency (ESA)'s Mars Express satellite.

Mars is often cited as the next destination in crewed space travel, after we conquered the moon fifty years ago. However, getting humans to Mars is a much tougher challenge than going to the moon. For a start, at their closest points Mars is 33.9 million miles (54.6 million km) from Earth, while the moon is 240,000 miles (384,000 km). Mars is more than a thousand times further away than the moon, so getting there would take a lot longer. Because of that, whoever went on that mission would have to be able to provide everything they need – oxygen, water, heat, shielding from space radiation, energy, fuel, food and waste disposal – from materials they took with them or ones found in space. There are teams of scientists working worldwide towards solving these issues, but when it comes to getting humans to Mars there is still a long way to go.

Jupiter

The closest of the gas giants, and the largest planet in our solar system, Jupiter shines brightly in our sky. When Jupiter is visible, it is usually the third brightest 'star' in the night sky, and it glows with a white hue. Jupiter can be seen at some time during the night on more than 300 nights of the year, so it is a good target if you are looking to find your first planet.

There is a big gap between Mars and Jupiter in the solar system, where the asteroid belt lies. Compared to Mars, Jupiter is more than three times further from the sun, so light takes forty-three minutes to reach it.

Jupiter is so big, it can fit all the other planets in the solar system, put together, inside it. It's a gas giant, which means it has no rocky surface like Earth, Mars, Mercury or Venus. Instead, it is a huge ball of mainly hydrogen and helium. Through a telescope, you can see the storms raging on Jupiter's surface, including its famous red spot. The Great Red Spot is a storm that has been going for hundreds of years, and it is so big that Earth could fit inside it three and a half times.

It is also possible to see some of Jupiter's moons using a telescope. Jupiter has over seventy moons and counting, four of which were famously discovered by Galileo in 1610 and are therefore named the Galilean moons. The satellites were the first moons in the solar system to be discovered orbiting another planet. Their names are Io, Europa, Ganymede and Callisto.

In close proximity to the heat given out by Jupiter, the moons are interesting worlds of their own. Europa's icy surface is thought to be hiding oceans of liquid water underneath, and this is one of the most promising places to look for life in our solar system. Likewise, Io, which is the innermost of Jupiter's four Galilean moons, is the most geologically active body in the solar system, which makes it a fascinating place to explore.

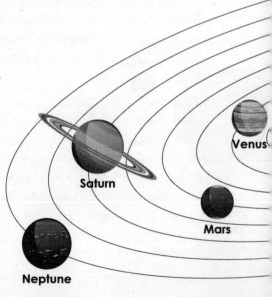

Saturn

Almost double the distance from Jupiter to the sun lies
Saturn, which sits 9.5 times further away than Earth.
Light takes seventy-nine minutes from the sun to reach
Saturn.

Just like Jupiter, Saturn is a gas giant with exciting moons.
Spotting Saturn with the naked eye is easy when it's in
opposition, but you won't see its signature rings without
looking through a telescope. Like Jupiter, Saturn can be
seen on more than 300 nights of the year.

Through a telescope, you can clearly see the ring system.
Personally, I believe everyone should see the rings of

Saturn through a telescope once in their lifetime. I will always remember that feeling I had when I saw them, those incredible rings that are so iconic, of which I had seen thousands of photographs. They were there, in real life, in real time. Of course, it was not quite real time, as light from Saturn takes about eighty minutes to reach us here on Earth. But it's the closest to real time I have seen those rings and it was magical.

Saturn is not the only planet with rings – all the gas giants have them. It's just that Saturn's rings are the most prominent. Exactly how the rings formed is not fully understood, but we know they are between 10 and 100 million years old, which is young for the solar system. It is thought that the rings were formed by shattered bits of asteroids, comets and moons, but a NASA photograph taken by the Cassini spacecraft in 2014 appears to show a moon forming from the rings, suggesting the rings and the moons are linked together.

Like Jupiter, Saturn has some really exciting moons. In 2017 molecular hydrogen was found on the surface of Enceladus, which also has an ocean of salt water under its surface, and this means the moon has all the ingredients it might need for life to exist.

Titan, another one of Saturn's moons, has weather that appears similar to the weather systems on Earth, except instead of water it rains methane. The icy world is so cold that its surface is home to oceans of liquid methane, and it has clouds which rain the hydrocarbon.

Uranus

While you cannot see Uranus from Earth with the naked eye, it can be spotted using a telescope. Uranus has a faint blue or green colour, caused by its atmosphere made up of mostly hydrogen and helium, with some helium sulphide and methane.

Like Saturn and Jupiter, Uranus has rings. Ten of them, in fact, that we know of currently. However, its rings loop around the planet from top to bottom rather than around its equator, like Saturn's rings. Another difference is that Uranus's rings are dark, unlike Saturn's brighter rings, which reflect more light and so can be seen easily through a telescope.

The third biggest planet in the solar system, Uranus's radius is four times the radius of Earth, and its average distance from the sun is 1.8 billion miles (2.9 billion km). This means light takes almost 160 minutes to reach Uranus from the sun, or two hours and forty minutes. One day on Uranus is seventeen hours but it takes eighty-four Earth years to complete a lap of the sun, making a year a very long time on Uranus.

Because of its distance from the sun, Uranus is an icy place, and the coldest temperatures measured there are the coldest of any planet in the solar system. The average temperature on Uranus is −216 degrees Celsius, which is even slightly colder than Neptune.

Neptune

Like Uranus, to spot Neptune from Earth you need access to a telescope. Neptune is the furthest planet from the sun, so receives 1,000 times less sunlight than we do here on Earth. Sunlight takes four hours to reach Neptune, and during that time the sunlight fades.

But it is far from a dull place. In fact, storms and extreme winds cover the planet, to the extent that astronomers still do not understand how the weather systems are created, since it receives so little light. Being the furthest from the sun makes Neptune one of the coldest planets, with average surface temperatures of −214 degrees Celsius, though it might not always have been so cold there. Astronomers think Neptune was created much closer to the sun, billions of years ago, before being expelled to where it now sits, at an average distance of thirty astronomical units, or 2.79 billion miles (4.49 billion km).

Neptune is the fourth largest planet in the solar system and its mass is seventeen times the mass of Earth. Neptune is not a solid planet, so a day on Neptune is measured by how quickly clouds at its equator take to make one complete rotation, which is eighteen hours.

Neptune has seven moons, one of which was first spotted in 2014 and was given the name Hippocamp at the start of 2019. The name comes from the genus seahorse, because the astronomer who named it is a fan of scuba diving. Hippocamp is a tiny moon, only 42 miles (68 km) in diameter.

The Dwarf Planets

Pluto cannot be seen with the naked eye, and it can be tricky to find with a telescope, too. It's become more and more faded since 1989, when it reached its closest point to the sun, and will continue to fade until the year 2114. It will be bright again in 2207. One for the great-great-grandkids, maybe.

Up until 2006, Pluto was considered the ninth planet in our solar system. However, the planet was demoted because it turned out to be too different from the others. The official definition of a planet, by the International Astronomical Union (IAU), has to meet three criteria: it has to orbit the sun, be big enough to have a nearly rounded shape due to its own gravity and to have cleared its orbital path of debris.

Pluto failed to meet the third criteria because it shares its orbit with asteroids and other dwarf planets. As they started to study the solar system in more detail, astronomers began to discover objects almost as big as Pluto. However, the reclassification of Pluto in 2006 was triggered by the discovery of Eris, a dwarf planet bigger than Pluto. The official term 'dwarf planet' was introduced in 2006, as a way to redefine Pluto without having to add more planets to the solar system.

There are five known dwarf planets in our solar system, according to the International Astronomical Union. There is Ceres, which sits in the asteroid belt between Mars and Jupiter, and then Pluto, Haumea, Makemake and Eris, which all lie much further away than Neptune, in a region of the solar system known as the Kuiper Belt.

Ceres, because it is much closer than the other dwarf planets, is the easiest dwarf planet to see in the sky. When it is in the right place, and the conditions are good, it can be seen with a pair of binoculars or a small telescope. Compared to the planets, Ceres is small – only 590 miles (950 km) in diameter. It takes 4.6 years to orbit the sun.

There are many other objects that could become classified as dwarf planets in our solar system. In fact, some estimate there could be hundreds of them out there.

2

HOW TO STARGAZE

Stargazing does not have to be complicated. It really can be as simple as looking up and noting the things you can see. If you look regularly enough you will start to notice patterns in the stars, planets moving and the moon changing its shape. That's stargazing. But once you start to try and search for new stars, constellations and objects in the night sky, there are a few tricks and techniques that will become very valuable to you.

How to use a star chart

There are plenty of amazing apps, both paid and free, available to download on your phone, tablet or laptop to help you understand what you can see in the night sky on any particular evening (see Further Resources, at the back of this book). However, as we know, using these during stargazing can prevent your eyes from fully adjusting and getting the best possible view of the night sky, because of the light they give off. This is why it can be useful to be old school and use star charts.

A star chart is essentially a map of the night sky that will allow you to navigate and find what you are looking at, depending on the time of year, where you are and what direction you are facing. You can buy a star chart, or planisphere, online, tailored for your specific latitude. To buy online, try searching for 'planisphere' rather than 'star chart', unless you are looking to buy a whiteboard with star stickers for your kids who are not so well behaved.

If you are based in the UK, which is at 55 degrees latitude, you would look for a chart that says 50 North or 50 N. These can be used in the UK, northern Europe and Canada. If you are in the northern half of the USA, get a 40–50 N chart; in the southern half, you will need 30–40 N; and so on. If you are uncertain what chart to get, search online for the latitude of your country and find a chart that is a close number to that.

At first glance, a star chart might seem overwhelming and complicated to use, but the idea is in fact very simple: it's a map of the night sky. Imagine there is a dome surrounding Earth on to which every star is painted, then imagine this dome is written on to a map. In the centre of the chart is the zenith, or the most northerly point in the sphere, which is where the North Star can be seen, also known as Polaris. As you rotate the star chart, you are mimicking what happens to the night sky as time passes and the Earth continues to spin. When it is aligned correctly, the star chart is what you should see when you look up. The edges of the chart are the horizon.

Some charts glow in the dark, while for others you might need a red light to be able to see what you are doing. Remember, it is important to find the darkest place you can get to – in a city this can be tough, but find an open park or walk to the top of a hill, go to the rooftop of a nearby skyscraper if you can – and give your eyes time to adjust.

To use the chart, first align it so the direction it is facing is the same one you are – if you are sat facing south, move the chart so the south position is pointing that way. Then you can start to find the brightest stars in the sky using the chart. It is usually a good idea to find easily recognisable constellations first, like Orion during winter in the northern hemisphere, and go from there.

The technique astronomers use to find other constellations once they know the direction they are facing and have found one constellation is called star hopping. There is no right way to star hop, so just use a method that seems easy to you. Find at least three stars you are familiar with, that you know you have identified

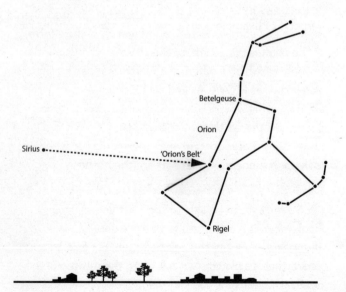

Southwest to west, April evenings

correctly. Then, you can create a pattern with these three stars to point you in the direction of another.

For example, find the three stars that make up Orion's belt. Going from right to left, look at those stars as if they make a straight line. Continue that straight line to the left and you will find a bright star. This is Sirius, or the dog star, and it is in the constellation of Canis Major. We will take a look at specific constellations later, but you get the idea.

Star hopping is usually best used to find interesting objects to look at through binoculars, ones that are more faint and harder to spot using the naked eye. On the star chart, it will tell you what you can see next to a fainter star, which you can use to star hop to find your target. For example, just south of the stars in Orion's belt is the Orion Nebula, which can be fascinating to look at through binoculars. It can be viewed with the naked eye, too, but looking at it more closely brings out the incredible detail of the network of gas and dust, which is in itself a star factory where baby stars are waiting to be born.

If you are spending a long time looking at the sky, remember to adjust your star chart to make up for the spinning of the Earth. After six hours, stars will have moved a quarter of the way around the sky.

If you find an incredibly bright star that is not in your chart, that is a planet. Note down where it was and the time, along with the size and colour, and you will be able to identify it. After a while, you will get to know how to identify each of the planets easily. Another telltale sign that it is a planet and not a star is if it traces the same path across the sky as the moon or the sun.

Once you have used your planisphere to identify some constellations, you will soon know a few of them off by heart. There are eighty-eight named constellations recognised by the IAU, but not all of them can be seen from one point on Earth on a given day, so you will become familiar with those that you can see.

In the northern hemisphere's night sky during winter, a popular star-hopping technique is to make a letter G out of the sky's seven brightest stars: Sirius, Procyon, Pollux, Capella, Aldebaran, Rigel and Betelgeuse. If Betelgeuse is left out, it can make a hexagon.

This letter G and the hexagon are examples of asterisms which are patterns in the sky that are not official constellations. Orion's Belt and the Plough, also known as the Big Dipper, are also examples of asterisms, both of which make up parts of official constellations.

In the southern hemisphere, a similar star hop is to look for the Southern Pointers, the two bright stars of Alpha and Beta Centauri – these are part of the constellation of Crux and point towards the south pole.

Stargazing with your eyes

Everyone knows that feeling in the middle of the night when you wake up and realise you need a wee. You lie there for a few minutes trying to decide between going back to sleep and putting it off until the morning (normally my preferred option) or just getting up and going. Eventually, it becomes so desperate that you must leave the warmth of your lovely bed and go to the cold loo.

You make it out of the bedroom just fine. You get to the bathroom and decide you need to see what you are doing. If you have to put the light on in the hallway or in the bathroom, though, when you come back to the dark room, suddenly you can't see anything at all. This is what it's like trying to look at the stars when your eyes are not adjusted to the darkness.

The point of this story is, your eyes need time to adjust to the darkness so they can see the light from the stars.

The first step in your eye adjusting to changes in light concerns the pupil. If you get someone to close their eyes for a few seconds, then open them again, you should notice their pupils becoming much smaller in reaction to the increased brightness. The pupil is a hole in your eye that lets light in. When it's darker, the pupils need to be bigger in order to let more of the light in.

This process happens in a matter of seconds, but it's not the only step in your eyes adjusting to the darkness, a process called dark adaption. In fact, the whole process takes forty minutes.

Behind your iris inside your eye is your lens, which, if it works correctly, bends the light so it focuses exactly on your retina, the surface at the back of the eye connected to the optic nerve. For many people, like me, the lens doesn't work as it should. In my case, my lenses bend the light in my eye way too much, focusing the light in front of my retina, so I am short-sighted.

On the retina there are light-sensitive little detectors called rods and cones. Rods are much more sensitive than cones, and they kick in when the light is low. It is these little detectors that take a while to adjust to the low light. In dimly lit situations, the rods start to produce a molecule called rhodopsin, which allows the retina to be stimulated when light hits it. This process can take an hour to happen fully but is ruined straight away if you look at a bright light.

If you are in your garden, on a balcony, roof or even just sticking your head out of your highest window, turn off all the lights in your house. If you are using a phone, install an app with a red-light filter and turn the brightness down as low as possible, so you can still look at the screen. Ideally, though, you want to put the phone away. Then, be patient. Around ten to fifteen minutes later, you should start to see a lot more. Then, forty minutes later, your eyes will be fully adjusted.

There is another trick related to looking at faint objects. Looking straight at something makes the light hit a part of the retina called the macula. This is where most of the 6 million cones in your eye sit. But there are 120 million rods in your eye, and these sit around the macula.

If you can't quite see a faint star, or can't see it very well, look slightly away from the target you are trying to look for. You will see it appear in the corner of your eye, as if by magic. This is because it's now being picked up by the rods around your macula, instead of the cones within it. This is called averted vision.

If you really need a torch to find your way around but don't want to ruin your adapted eyes, use a red light. You can buy red lights online or use a red bicycle light if you have one. The red part of the electromagnetic spectrum is what the rods in your eyes are least sensitive to, so it's the best light to use if you do not want to ruin the dark adaption you spent an hour on. Having said that, try not to get a light that is too bright, as even red lights will ruin dark adaption if they are bright enough.

It really is best not to use any source of light at all.

How to stargaze with equipment

When you first start stargazing, it is best not to buy any equipment because you will be surprised how much you can see with the naked eye. Once you have become used to finding your way around the night sky, the first piece of equipment you should buy is a pair of binoculars to help you see even more in the night sky.

Binoculars

Binoculars are described according to their magnification and aperture. This means a pair of 7 x 50 binoculars makes objects seven times bigger and has an aperture of 50 mm, which means the main lens is 50 mm in diameter.

For a starting pair of binoculars, it is best to avoid the highest magnification, 12 or higher, because these will be large binoculars that are difficult to hold steady. However, a magnification that is too low will also cause problems, because the circular beam of light leaving the binoculars, called the exit pupil, could become larger than your own pupils. If this happened, you would not see the image correctly.

Younger people with healthy eyes tend to have pupils between 2 and 7 mm, but for older people the dilated pupil size decreases to around 5 mm. You can measure the 'exit pupil' of a pair of binoculars or calculate it by dividing the diameter by the magnification. The maximum exit pupil you should aim for is seven.

To help keep your binoculars steady, attach them to a tripod. If you find this kind of set-up uncomfortable, you can also buy a parallelogram binocular mount, which bolts on to the tripod and means it is much easier to look through the binoculars.

Telescopes

Once you have become comfortable stargazing with binoculars, the next step is to buy your own telescope. Some people even build their own telescopes. The important thing to remember here is not to rush into buying a telescope. The cheapest telescopes do little more than a pair of binoculars, and the most expensive ones can be incredibly expensive. Only once you are sure you would like to invest in a telescope should you go and buy one.

There are two types of telescope – refracting and reflecting. The names are given depending on the way they magnify the light. The first kind of telescope to be invented was a refracting telescope, which uses lenses to magnify light. The problem with the first refracting telescopes was that they experienced chromatic aberration. Light of different wavelengths is bent to a different extent when it goes through a lens. This creates a multicoloured halo of light around the image. Today, refracting telescopes use more than one lens to correct for this effect, but that makes them more expensive than the alternative.

The other type of telescope is a reflecting telescope, which uses mirrors to collect the incoming light into an image. There are also combinations of refracting and reflecting telescopes, known as catadioptric telescopes.

This book will not teach you what kind of telescope to buy, because it is aimed at those starting out on their journey into astronomy. To experience stargazing through a telescope, visit your local observatory, or alternatively, there are many resources in both books and online that will explain in detail what kind of telescope is right for you.

Astrophotography

If you are interested in taking photographs of what you see, you might not need to invest in an expensive set-up. If you have a phone with a good camera, you can take simple photos of the stars; for example, the constellation of Orion can be caught on camera using a recent phone. Phones are great for capturing aurorae and planets, too, and there are a lot of apps aimed at helping people use their phones for astrophotography, detailed in the Further Resources section.

Using a camera, a great way to start capturing photographs of the night sky is by looking at star trails.

As the Earth spins, the stars appear to make a pattern through the sky. If you have a decent camera and a tripod, you can point it at the sky and set it on a very low shutter speed. If the shutter is open for an hour, you should get an incredible pattern from the star trails in the sky. A good idea is to make sure the aperture is as small as possible and set the ISO to around 200. Make sure to operate the shutter remotely so that there is no shake as the camera starts to capture the image.

Taking photographs with exposures of roughly less than thirty seconds will allow you to take photographs that show the stars as bright points, instead of streaks. This means you can capture the constellations as you see them. The longer your focal length is, the shorter the exposure time you will need. It is possible to capture the Milky Way on camera, too.

The next step is mounting your camera to a tracking mount which tracks the movement of the night sky. This allows you to take multiple, long exposure images of the same object and stitch them together to get the resolution you are looking for. If you know what you are doing, some amazing images can be taken just using an entry-level DSLR or mirrorless camera and a tracking mount. These mounts are not cheap, but they can make or break your astrophotography. To track the stars accurately, the mount needs to be sturdy and reliable.

The next step, if you want to take images of deep-sky objects like galaxies or nebulae, is to attach a telescope to the front of your camera, which is attached to the tracking mount. The smaller telescopes that are easy to carry around can be quite expensive, and if you are just looking to do back-garden astrophotography you can mount your camera to a larger telescope.

3

WHAT TO SEE
– THE STARS

There are about 5,000 stars visible to the naked eye from Earth, and from any one point on Earth, this number is halved because we can only see half of the night sky. Although they look like tiny points of light, stars – like our sun – are huge balls of gas and plasma with temperatures so high they are capable of fusing elements together to create new ones.

The stars themselves appear to be moving through the sky throughout the night, but the movement is actually our Earth rotating on its axis and orbiting the sun. To describe where stars appear in relation to the Earth, we use an imaginary sphere covering Earth called the celestial sphere. Each star has a position on this sphere, and the celestial north and south poles are directly above Earth's north and south poles.

How the stars move

Did you know that the stars change night by night? If you looked up at 7 p.m. one evening, and at the same time again the very next day, the sky would by slightly different – by four degrees, to be precise. This is because it does not take the Earth exactly twenty-four hours to fully rotate; it actually takes twenty-three hours and fifty-six minutes, which is known as a sidereal day. This is due to a phenomenon called precession. Because of this, if you looked at the sky at 7 p.m. one evening, it would be exactly the same the next day at 6.56 p.m., four minutes earlier. Because of this change, we start to slowly see different constellations in the sky and the summer night sky is totally different from the winter night sky.

The Earth rotates on its axis, but the North and South Poles are always facing the same direction. Because of this, constellations and stars near the celestial North Pole are always visible from the northern hemisphere and likewise for the celestial South Pole and the southern hemisphere. Some constellations can be seen only from certain points, while other parts of Earth see the same constellations but at different times of the year. For example, in the northern hemisphere you will never see the asterism the Southern Cross, and the North Star cannot be seen from the southern hemisphere. Orion, however, passes through both.

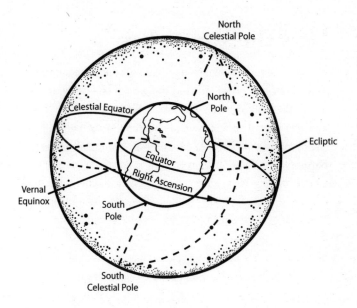

The way the Earth orbits the sun also plays into what you can see. The Earth is at a slight tilt of 23.5 degrees, compared to the plane in which it orbits around the sun. This is why we have seasons – winter in the northern hemisphere is when the northern hemisphere is tilted away from the sun, which means solar radiation reaching Earth is less intense, so it is colder. Summer is when it is tilted towards. This is why, during summer, the Arctic Circle experiences twenty-four hours of sunlight, and it is plunged into twenty-four hours of darkness in the winter.

On 20 or 21 June, the northern half of Earth is at its maximum tilt towards the sun, and that gives us the longest day of the year in the northern hemisphere, known as the summer solstice. Around 22 or 23 September, the Earth is halfway round the sun: the autumnal equinox, where day and night are the same length as each other, around the world. The point at which the northern hemisphere is tilted away from the sun, the winter solstice – the shortest day of the year – happens around 21 or 22 December. The spring equinox happens on 20 or 21 March.

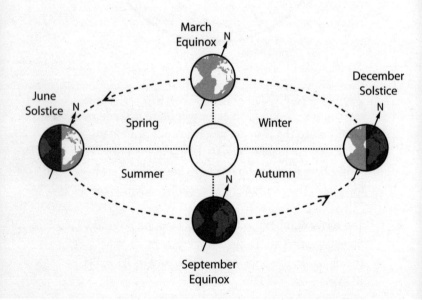

The celestial sphere

As you look into the night sky, astronomers imagine every star you see lies on a gigantic sphere that extends around our whole planet, which rotates slowly. In reality, the stars you see lying next to each other in the sky are not necessarily close together in the galaxy. The celestial sphere looks as if it rotates around an axis that runs through the Earth's North Pole and South Pole, from east to west. This is because the Earth is rotating from west to east during the day and night.

The altitude of any object in the sky is its position above the horizon, in degrees. An object on the horizon has an altitude of 0, and an object directly above the observer's head, a point called the zenith, is 90. If an object has a negative altitude, it is below the horizon. The altitudes are given relative to the position the observer is in, which means they vary depending on where you are in the world.

The other number used for describing the position of objects is azimuth. This is given in a number relative to the direction of north – north is 0, east is 90, south is 180 and west is 270 degrees. The meridian is a line which runs along the celestial sphere, from the position where altitude and azimuth are zero, known as the north point, through the zenith, the south point, the nadir and back to the north point. The imaginary boundary between north and south is the celestial equator.

The zenith at the North Pole is the north celestial pole, and this is the point that never changes as the Earth rotates. The North Star, also known as Polaris, is extremely close to the north celestial pole, so it never moves in the sky in the northern hemisphere.

To measure the number of degrees between one object and another in the sky, you can use your hand. Roughly, when you stretch your arm out fully, the width of your little finger is the width of one degree, the clenched fist

THE NORTH STAR

The North Star, also known as Polaris, is extremely close to the north celestial pole, so it never moves in the sky in the northern hemisphere. Its altitude in the sky is the same as the latitude of the place you are viewing it from, for example:

London: 51 degrees

New York : 41 degrees

San Francisco: 38 degrees

Paris: 49 degrees

Tokyo: 35 degrees

Hong Kong: 22 degrees

is ten and a fully stretched hand is twenty. This can be helpful when you are trying to star hop and know the relative positions in the sky of two objects.

Light years are a measure of distances, but there are a couple of other terms used for dealing with incredibly huge distances in space. One is called a parsec. A parsec is equivalent to 3.26 light years, but it is not just a random number. It is determined by something called apparent shift. If you hold your hand out at arm's length, with one finger up, looking through only one eye, then close that eye and open the other, you will see the position of your finger appears to change. The same thing happens with stars in the night sky when the Earth moves, from one side of its orbit of the sun to the other. This apparent shift is called parallax shift. The distance a star has to be for the parallax shift to be 1 arc second, which is 1/3,600ths of a degree in the sky, is a parsec. This is an important technique for calculating the distance between Earth and stars.

Magnitudes

Whether it is a planet or a star, how bright an object appears in the night sky makes a big difference to whether or not you might be able to see it on a given night with the naked eye or if you'll need binoculars. This is why it's important to understand the scale astronomers use to describe the brightness of objects in the night sky.

Some stars are really bright but really far away, while others that are less bright, but closer, may appear exactly the same brightness in our sky. This is why astronomers use two measures of how bright a star is – absolute magnitude and apparent magnitude. The apparent magnitude is how bright a star appears in the sky. Absolute magnitude is the actual amount of light that is given off by the star, and it is defined as the magnitude the object would have if it were being viewed from a set distance, ten parsecs, away.

Apparent magnitude is what matters to people who want to stargaze, so we will refer to it just as 'magnitude' from now on. It is measured on a logarithmic scale, and lower numbers are brighter, which means a first-magnitude star is about 2.5 times brighter than a second-magnitude star and 100 times brighter than a sixth-magnitude star. The sun has a magnitude of −26.7, the full moon is −11 and the brightest star, Sirius, is magnitude −1.46.

MAGNITUDES

The brightness, or magnitude, of a night-sky object is measured in a slightly confusing way – lower numbers are brighter. Here's a rough idea of how bright different objects are in the sky:

The sun: −26.7	Sirius: −1.46
Full moon: −11	Betelgeuse: 0.42
Venus: −4.6	Andromeda
Mars: −3	Galaxy: 3.44

For a rough guide, in a dark sky area with good conditions you are likely to be able to see stars with magnitude 6 or 6.5. With slightly light-polluted areas you are likely to be able to see magnitude of around 4.5. There is a test that can be done during winter in the northern hemisphere, using the constellation of the Great Square of Pegasus, which we will go through later.

The planets vary in magnitude, but the maximum magnitude of Venus is −4.6, Mars is −3.0, Jupiter is

–2.7, Mercury is –2.0 and Saturn is –0.5. Uranus, at maximum 5.7, and Neptune, at 7.7, are unlikely to be spotted with the naked eye.

Using binoculars, you can increase the range of magnitudes you will be able to see, but the extent depends on the particular binoculars. A pair of binoculars with an aperture of 40 with increase the magnitude you can see by +4.4, for example.

Twinkles

The twinkle in a star is nothing to do with the star itself, instead it is what happens to the light as it passes through Earth's atmosphere. The light is refracted in different directions, making the star look as though it is changing in brightness. Stars appear to twinkle more than planets because they look like points of light rather than disks. The light from planets is spread over the disk, and one side might look as if it moves in one way, but the other side will balance it out. When planets are lower in the sky, near the horizon, they might appear to twinkle because there is more atmosphere in the way between you and the planet.

Types of stars

All stars have one thing in common – at some point in their lives they are home to nuclear-fusion reactions. This means in the core of the star, where there is immense heat and lots of energy, atoms are fusing together to form larger elements.

The basic reaction for fusion is two atoms of hydrogen combining to form helium. The total mass left over after the reaction is slightly less than the mass we started with, and the difference is converted to energy, using the famous formula $E=mc^2$ – where E is the energy given out, m is the difference in mass and c is the speed of light. Because, as we know, the speed of light is such a huge number, it takes only a tiny difference in mass for a lot of energy to be given off. This is the process that powers stars.

The star will fuse hydrogen into helium and heat up. As it heats up, eventually helium atoms start to fuse together, creating even heavier elements. We know that in most stars, every element up to iron in the periodic table is created this way. In fact, all of the elements in the universe up until iron were made in the core of stars. The carbon in your bones, the oxygen you breathe and the sodium you eat were all created in the heart of stars.

Stars come in all sorts of shapes and sizes, and a few of the main types of stars are listed below.

Protostar

Before a star becomes a star, it is a protostar. A protostar forms as a result of a huge collection of gas, mainly hydrogen, that slowly stars to clump together through the force of gravity. Eventually it will collapse down into a spherical shape, in a process that takes around 100,000 years.

T Tauri star

Once a protostar starts to hold together under gravity, the pressure from this gravity becomes a source of energy and heats up the star. But there is a stage of around 100 million years during a star's lifetime when it does not have enough energy yet to start nuclear fusion yet it is big and bright, so looks like other stars in the sky. These are called T Tauri stars.

Main sequence

Most of the stars in the universe fall into a category called main-sequence stars. The size and brightness of a star will depend on the amount of gas that collected to form the star in the first place, and main-sequence stars can vary in brightness, mass and size.

Stars on the main sequence can vary from 0.8 times the mass of the sun, to 100 times the mass of the sun. What they have in common, though, is they are all generating fusion reactions from hydrogen into helium and producing a lot of energy. This energy creates a force that pushes outwards, which balances the gravitational forces pushing in on the star, creating what is called hydrostatic equilibrium. This will continue until the star uses up all of its core hydrogen. Most stars on the main sequence are dwarf stars.

Below are some of the most common dwarf stars you can see:

RED DWARF

One type of main-sequence star that is much cooler than our sun is a red dwarf. In a red dwarf, fusion is going on, but it happens so slowly that it takes much longer for the star to run out of fuel compared to other main-sequence stars. This means they can burn for billions, if not trillions, of years. Because of this, they are the most common type of star in the universe. For example, Proxima Centauri is a red dwarf.

YELLOW DWARF

Stars with similar masses to our sun are called yellow dwarfs. They are much hotter than red and white dwarfs.

WHITE DWARF

Once a red giant has run out of fuel, and no longer has enough mass to provide pressure to fuse its elements together, fusion stops. The star no longer has the outward pressure to balance out the gravitational forces and the star will collapse. It will continue to shine, but fusion is no longer happening in a white dwarf, which spends the rest of its days cooling down, over hundreds of billions of years.

Red Giants

For many main-sequence stars, like our sun, once all the hydrogen in the core has been used up, fusion will stop in the core and start in a shell around the core. This causes the star to grow in size, up to 100 times the size it was as a main-sequence star, and glow red so it becomes a red giant. When the hydrogen is used up, it starts to fuse helium and heavier elements. If a star is between eight and twenty-five times the mass of the sun, it will be able to fuse elements as heavy as iron. This stage will last a few hundred million years.

Supernovae

If a star is between around ten and twenty-nine times the mass of the sun, it will not collapse to form a white dwarf. Instead, the star will die in a dramatic explosion called a supernova. What is left is an incredibly dense type of star called a neutron star, which can be tiny, around 20 km in diameter, but have an incredibly huge mass, more than double that of the sun. One teaspoon of neutron-star material weighs about a billion tons, which means gravity on the surface of a neutron star is two billion times the gravity we feel on Earth.

In 2017, a collaboration of astronomers around the world measured the signals given off by a pair of neutron stars colliding, in the first measurement of gravitational waves, which opened up a new window into how to study the universe and gave us evidence of a black hole being created for the first time. These waves, which are ripples in space–time created by such high-energy events, were first predicted by Einstein 100 years before we ever managed to find them.

Neutron stars rotate very rapidly, as fast as 43,000 times a minute. Some of them can be identified because they shine out huge amounts of radiation from two ends, which we can see as a pulse of radiation, like an intergalactic lighthouse, flashing between 0.1 and 60 times every second. These are called pulsars.

If a star starts out greater than twenty-nine times the mass of the sun, it will also undergo a supernova explosion but will end up as a black hole instead of a neutron star. This type of supernova, whether there is a black hole or a neutron star created at the end, is called Type II.

Supergiants

Stars greater than 100 times the mass of the sun, called supergiants, explode and leave no trace at the end of their lives. They use a lot of fuel, which means they have relatively short lifetimes compared to other stars: only a few million years.

Multiple-star systems

Binary stars are systems in which there are two stars orbiting around each other. When there are three or more involved in the set-up, these are known as multiple-star systems. The majority of the 'stars' you see in the night sky are in fact two or more stars orbiting together.

There are two general types of binary stars – wide and close – which, as you can imagine, depend on how far away the stars are orbiting each other. Close binaries will often end up in a situation where one star begins to consume material from its companion star, and even pull it in completely. When this happens, there is an explosion called a Type I supernova.

Visual binaries mean the two stars are far enough apart that, through a telescope, both can be seen as separate stars. The very first binary star system discovered was a visual binary, observed by Galileo in 1617, when he looked at the second star along from the handle in the asterism the Plough.

We have also discovered planets orbiting binary or multiple-star systems, which means the planets have two or three suns.

The closest stars to Earth form a binary system called Alpha Centauri, known as Alpha Centauri A and B. Near this star system there is Proxima Centauri, so often they are referred to as a multiple-star system.

Cepheid variables

A Cepheid variable is a kind of star that changes its brightness, varying in temperature and size as it does so. They are bright but distant stars, and they are useful for measuring distances. This is because the absolute magnitude of a Cepheid variable can be determined by the period of its pulses. This means we know its brightness from one parsec away, and we know the brightness we see on Earth, so we can calculate the distance that star sits from Earth.

SEASONAL SKIES

The later pages in this chapter go into great detail about individual constellations, but first we need to get to know what is visible during each season in the two hemispheres. Of course, the planets will be doing their own thing, so make sure to keep an eye on them, too.

Northern hemisphere

In the northern hemisphere, anywhere north of 40 degrees, the constellation Ursa Major, which contains the handle known as the Plough or the Big Dipper, can always be seen. Draco, Cassiopeia, Cepheus, Ursa Minor – which contains the North Star or Polaris – and Lynx are all circumpolar, too, which means they can always be seen, and they never set below the horizon.

Things you can spot in the northern hemisphere during different seasons:

WINTER

If you have ever looked up during a winter's night in the northern hemisphere, you will be familiar with Orion. The bright constellation is easy to recognise, and it is a great place to start as the direction of the belt can point you in the direction of others. For example, if you follow the belt from right to left and keep going in that direction, you will reach a bright star, Sirius, which is the brightest in the sky. Sirius makes up part of Canis Major, another winter constellation.

Elsewhere in the sky you will be able to find Taurus, Auriga, Hyades, the Pleiades, Perseus, Cassopeia and Gemini, with its two bright stars, Castor and Pollux. The Milky Way can be seen across the sky, from Cassiopeia to Orion.

The Andromeda Galaxy can be spotted during winter, too, using the constellation of Pegasus.

SPRING

During spring in the northern hemisphere, recognisable constellations include Leo, Hydra and Bootes. Leo is the most distinctive of these, and just west of Leo you can find Cancer, which is the least bright of all the zodiac constellations.

The Lyrid meteor shower normally peaks at the end of April, in the Lyra constellation.

The spring is also when a deep-sky object called the Beehive Cluster appears in the skies, and it can be seen with the naked eye when the sky is dark enough; but when it is viewed through a pair of binoculars, dozens of individual stars can be made out. It is in the constellation of Cancer.

SUMMER

Due to the lighter nights, stargazing in summer is not as easy as it is in winter. However, the weather can often be more favourable.

During summer in the northern hemisphere, the densest part of the Milky Way becomes visible, through which Cygnus can be seen in the centre.

A famous asterism of three bright stars, called the Summer Triangle, is formed of Deneb, Vega and Altair. The constellations Lyra, Cygnus, Aquila and Deplhinus are also distinct and easily recognisable summer constellations in the northern hemisphere. Hercules and Libra can also be spotted during the summer.

AUTUMN

In autumn, the nights begin to darken again, and the Summer Triangle falls below the horizon. In the sky you can find the Great Square of Pegasus, Andromeda, Aries, Pisces, Aquarius and Perseus. But Pegasus and Andromeda dominate the sky with their distinctive shapes.

Autumn is a great time for meteor showers, with the Draconids, the Orionids, Taurids and the Leonids.

Southern hemisphere

If you are used to stargazing in the northern hemisphere, you will find that the night sky in the southern hemisphere is totally different from what you are used to seeing. This is because the southern hemisphere is looking out into the flat plane of the Milky Way, and there are fewer bright stars in the foreground in the southern hemisphere compared to the northern. It will also be an amazing experience, as the view of the Milky Way is incredible.

The North Star, or Polaris, is never visible in the southern hemisphere, and some of the constant features in the north – Cassiopeia, Ursa Major, Draco and Cepheus – are seasonal instead. But from the southern hemisphere you can see our closest neighbouring stars, dwarf galaxies and clusters of stars never visible from the north.

In the southern hemisphere, Carina, Centaurus and Crux are circumpolar, so they can always be seen.

Things you can spot in the southern hemisphere during different seasons:

WINTER
The Summer Triangle mentioned earlier becomes the Winter Triangle in the southern hemisphere, and it is upside down compared to how it looks in the north, too.

It's a good time to spot the nearest star system to Earth, Alpha Centauri. Although this consists of three stars, it looks like a single point with the naked eye. Using a telescope, all three can be seen. Alpha Centauri can be seen all year round in the south.

Look out for the Southern Cross, which is an asterism in the constellation of Crux. Because Crux can always be seen, it has been used for years as a navigation aid, like the North Star in the northern hemisphere.

If you enjoy looking at Andromeda in the north, you can find two other nearby galaxies with the naked eye in the southern hemisphere – the Large Magellanic Cloud and the Small Magellanic Cloud.

SPRING

In spring, you can spot Andromeda, Aquarius, Capricorn, Pegasus and Pisces – many of the northern hemisphere's autumn constellations.

The Southern Cross starts to be barely visible, just above the horizon, and the Milky Way moves towards the west during spring.

SUMMER

During summer, Orion appears low in the sky and points towards Rigel, with Sirius overhead.

The second brightest star in the sky is Canopus, which is often called the Great Star of the South and can be seen most of the year.

AUTUMN
As in spring in the north, Bootes, Cancer, Crater, Hydra, Leo and Virgo become visible during autumn in the southern hemisphere.

Star names

Before we start to talk about specific stars in the sky, it is important to note the way stars are named as a lot of them have at least two names. If a star in the sky is bright enough to have been recognised with the naked eye, it is likely to have been given its own name. This is called its proper modern name, as recognised by the International Astronomical Union, and many of these names were given to the star by ancient Greek or Arabic astronomers. There are 330 stars with proper names, according to the IAU, as of June 2018.

Other star names are simply derived from the constellation that particular star belongs to and a letter from the Greek alphabet. This is called a star's designation. The order does not necessarily represent the brightness of the star, but it sometimes does. For example, the brightest star in the constellation of Crux is called Beta Crucis, the second brightest is Alpha Crucis, and so on. Polaris, the brightest star in Ursa Minor, is also called Alpha Ursae Minoris. These designations can also be abbreviated, to Alpha UMi or α UMi.

In stargazing, the convention is to use the proper names for stars, and these will be the ones we are most likely to be dealing with. However, there are some occasions in the following chapters when we have used the designation for a star in the instances where a star does not have a proper modern name.

Likewise, for galaxies and nebulae, astronomers give names which are a series of letters and numbers representing how the object was discovered and what it is. The important ones also have proper names.

Star patterns

Asterisms

There are eighty-eight official constellations in the night sky, according to the International Astronomical Union. However, some groups of stars form easily recognisable shapes that are not official constellations. These are called asterisms.

Some asterisms, like Orion's Belt and the Plough, make up a small section of whole constellations. Others, like the northern hemisphere's Summer Triangle, are composed of stars from different constellations. The Summer Triangle is made up of the brightest stars from the Aquila, Cygnus and Lyra constellations, for example.

Often, finding asterisms can help with finding other constellations, so it can be important to learn these too. For example, to find the Summer Triangle in the northern hemisphere, look directly overhead during summer at around midnight. The most easily recognisable star is Vega, which shines blue-white. From there, look to the lower right to find Altair. The third point in the triangle is Deneb. From these three stars you can already identify three separate constellations.

STARS TO LOOK OUT FOR

Polaris

The North Star, Pole Star or Polaris is a very important star in the northern hemisphere as it shines above the North Pole, at the celestial North Pole. For this reason, whenever you can locate the North Star you can find out which way north is. If you face Polaris and stretch your arms out to your sides, your right arm will be pointing east and left arm will be pointing west.

The star does not lie exactly above the pole, though. Polaris lies less than one degree from the true North Pole, so if you take a photograph of star trails you will see Polaris trace a tiny circle around the pole.

The elevation of Polaris above the northern hemisphere is the same as the latitude of the person looking at it. So, observing Polaris from London, which has a latitude of 51 degrees, Polaris can always be found 51 degrees above the horizon. In Paris, it's 49; in New York, 41; and in Tokyo it is at 35 degrees. This means that measuring the elevation of Polaris can tell you what your latitude is, if you are in the northern hemisphere. But you have to find it first.

Although the North Star is not the brightest in the sky (it is around the forty-sixth-brightest star), it can always be found in the northern hemisphere. Even during a full moon, with an apparent magnitude of 1.97, Polaris is bright enough to see with the naked eye, so it is a reliable star.

HOW TO SEE IT

To find Polaris, you need to locate the constellation Ursa Major, which includes the asterism the Big Dipper, or the Plough. The Big Dipper or Plough is always moving, circling around the North Star, but it has two stars which always point towards it, which are the pointer stars called Dubhe and Merak. These are the two stars furthest away from the tail of the Plough, and a line from Merak to Dubhe, extended around five times the distance between them, points towards Polaris.

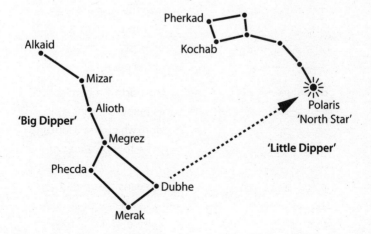

Polaris belongs to the constellation Ursa Minor, or the Little Bear, which is also known as the Little Dipper in the USA. It is not a very distinctive constellation, as it is made up of faint stars, but its proximity to Ursa Major makes it easy to spot. Its shape is similar to the Plough, with a long tail and a square at the end. Polaris marks the end of the tail of Ursa Minor.

HISTORY OF POLARIS
The further north you travel, the higher Polaris climbs in the sky, until you are at the North Pole and it is directly overhead. This hasn't always been the case because of the movement of Earth. When the pyramids were built in ancient Egypt, the North Star was another star called Thuban, which is in the constellation of Draco.

Polaris has had many names over the centuries. In the tenth century, Polaris was known as *scip-steorra*, which means ship star, in Anglo-Saxon England, because it was useful for navigation. In medieval times, it was known as *stella maris*, which means star of the sea. During the Renaissance, the name Polaris was first used in reference to the star, although back then it was a few degrees from the celestial pole.

Although the Earth's axis is shifting, Polaris will remain our North Star for hundreds of years. In fact, it is getting more aligned to the pole as the years go by. By March 2100, Polaris will be exactly above the north pole, which means if someone were to take a photograph of a star trail then, the star would not make a circle at all.

TYPE OF STAR

Polaris is in fact three stars, a multiple-star system made up of Polaris A, Ab and B:

- Polaris A is the biggest of the three, and it is a supergiant star six times heavier than our sun.

- Polaris Ab and B are both dwarf stars, but are just as hot as Polaris A.

- Polaris Ab is 2 billion miles from Polaris A

- Polaris B is 240 billion miles from Polaris A

The system of Polaris is a Cepheid variable, which is handy for astronomers to figure out distances. It is estimated that Polaris is 430 light years from Earth, which is not that far in astronomical terms. Because of its distance, however, it is thought to be as bright as 2,500 suns.

The Southern Cross

There is no single star in the southern hemisphere as useful as the North Star for navigation, as there aren't any stars near the celestial South Pole. It will remain this way until the year 4200 when a star called Gamma Chamaeleontis will pass close to the celestial South Pole. However, those in the southern hemisphere can locate the direction of south using the Southern Cross.

The Southern Cross is an asterism that forms part of the constellation Crux, the smallest of the eighty-eight official constellations. The Southern Cross is a group of four main stars – Alpha, Beta, Gamma and Delta Crucis, and a fainter fifth star, Epsilon Crucis – which is easy to recognise as each of the four stars has a magnitude of 2.8.

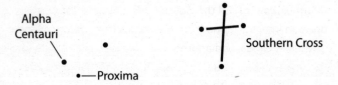

How to see it

To find the Southern Cross, there are two pointer stars called Hadar and Rigel Kentaurus, in the constellation of Centauraus. Rigel Kentaurus is also known as Alpha Centauri, the closest star system to Earth, which we learned about earlier, and Hadar is also known as Beta Centauri.

Going from Alpha Centauri to Beta Centauri, follow this line and you will reach the group of four stars that make up the Southern Cross.

When looking for the Southern Cross, there is a nearby set of four stars in the constellation Vela, known as the False Cross, which can be easily mistaken as the Southern Cross. To make sure you have spotted the Southern Cross, look for the fifth star, just below the right arm of the cross.

If you take the right arm and the top of the cross, these two stars are the pointer stars. Going from the top to the bottom, and following this line, the direction it is pointing is towards the celestial South Pole. If you extended the distance between the stars 4.5 times, you would reach the South Celestial Pole, and that direction is south.

Another way to find the South Celestial Pole is if you drew an imaginary line from the top to the bottom of the Southern Cross, and another line between the middle of the pointer stars to a star called Achernar, the place they intersect is the south celestial pole.

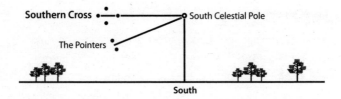

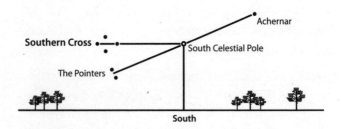

History of the Southern Cross

The Southern Cross was visible in parts of the northern hemisphere around 5,000 years ago, and it is thought the European explorer Amerigo Vespucci was the first to spot these stars in 1501, calling them the Four Stars.

NOTABLE STARS

Sirius

The brightest star in the night sky, Sirius has a magnitude of –1.46, almost twice as bright as the second brightest star, Canopus. Also known as the Dog Star, Sirius is part of the constellation Canis Major, or the Big Dog. There are two dog constellations – Canis Major and Canis Minor – and both are said to be following Orion, the hunter, like loyal canine companions. Whoever first came up with the idea Canis Minor looks like a dog deserves a medal for imagination, as it is only made up of two stars.

Like most stars we can see with the naked eye, Sirius is a binary star system made up of Sirius A and B. They orbit each other at a distance of between 8.2 and 31.5 astronomical units (a quick reminder: one astronomical unit is the distance between Earth and the sun).

Sirius A is a main-sequence star, about twice as big and twenty-five times as bright as our sun. Compared to other bright stars in the night sky, however, Sirius is quite a dim star in absolute terms. The only reason it appears so bright in our sky is that it is nearby.

Sirius lies just over eight light years from Earth, making it one of our closest galactic neighbours. It is moving closer, too. In the next tens of thousands of years, Sirius will gradually appear brighter and brighter in the night sky.

Sirius B used up its hydrogen a long time ago, turning into a red giant, and it is now a white dwarf.

HOW TO SEE IT
Finding Sirius is easy: just look for the three stars that make up Orion's Belt, going from right to left, and follow that imaginary line until you see a really bright star. That's Sirius.

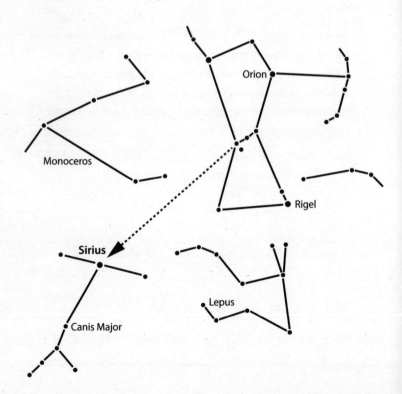

The history of Sirius

Because of its prominence in the sky, Sirius has long played an important role for humans all over the world. In Ancient Egypt, when Sirius rose slightly above the horizon, it marked the flooding of the Nile. The Egyptians worshipped the star, as the goddess Sopdet. The rising of Sirius in the summer marked the 'dog days' for the ancient Greeks, the time when heat and drought would wilt the plants. In the southern hemisphere, the star is present in winter. Because of this, the appearance of Sirius marked winter for the Polynesians, and it was used for navigating around the Pacific.

During summer in the southern hemisphere, Sirius can be seen early in the morning, when it rises before the sun, and in the evening when it sets after the sun. In the northern hemisphere, Sirius appears during winter, until mid-spring.

Canopus

The second brightest star in the sky, Canopus, lies in the constellation of Carina. In the northern hemisphere, it is hard to see as it never rises above 37 degrees latitude, which means those in the UK and most of North America will not see Canopus from home.

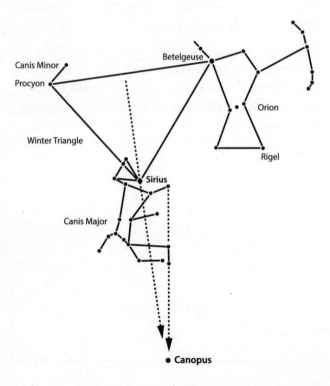

HOW TO SEE IT

It is easy to find during winter evenings for those who are south of 37 degrees, and in the southern hemisphere it can be seen for most of the year. During summer months in the southern hemisphere, Canopus is seen high overhead, along with Sirius.

To find Canopus in the northern hemisphere, first you have to locate the Winter Triangle. This is an asterism formed from Sirius, Betelgeuse and another star called Procyon. The same set of stars is the Southern Summer Triangle in the southern hemisphere.

Imagine you draw a line between Procyon and Betelgeuse, then from this line draw another line to Sirius. Extend the line away from Sirius, and the line will point to Canopus. You do not have to be incredibly accurate with this, because if Canopus can be seen where you are, you will not be able to miss it. It is often referred to as the bright star below Sirius, in the northern hemisphere.

Alpha Centauri

The closest stars to Earth are a binary system called Alpha Centauri, the stars are known as Alpha Centauri A and B. Nearby, there is Proxima Centauri, so often they are referred to as a multiple-star system.

HOW TO SEE THEM
The three stars can only be seen as individual stars using a telescope, but they are fun to spot because they are our closest neighbours. With the naked eye, they appear as a single point and are the third-brightest visible star in the night sky.

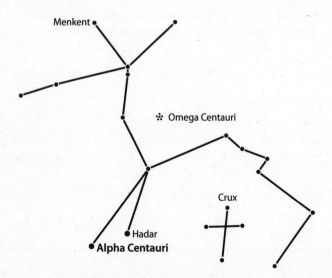

Any latitudes higher than 29 degrees north will never see Alpha Centauri, so you may have to venture to the southern hemisphere to get a glimpse. If you do, you are almost guaranteed to see it because it's circumpolar, so you can view it all year round. It makes up part of the constellation Centaurus, the centaur. Alpha Centauri makes the left toe of the centaur in the constellation. It is also near the Southern Cross, part of the Crux constellation.

Arcturus

The fourth-brightest star in the night sky belongs to the constellation Bootes, and it is named Arcturus, or Alpha Bootis. It is a red giant around seven billion years old. This means it has used up its inner hydrogen reserves and is now fusing heavier elements together. It is also relatively close to Earth, at around thirty-seven light years away.

HOW TO SEE IT
To find Arcturus, you have to first identify the Plough, or the Big Dipper. The asterism is made up of two parts – a handle and a bowl. If you look from the bowl to the handle, then follow the imaginary line the handle would

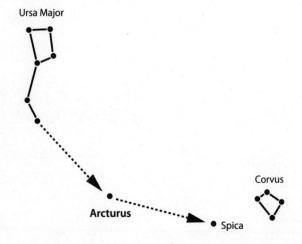

Ursa Major

Corvus

Arcturus

Spica

make if it continued through space. This will point you to Arcturus.

Arcturus marks the lower part of the Bootes constellation, which is most prominent in the spring in the northern hemisphere, which is autumn in the southern hemisphere. It can be seen from both hemispheres during this time. Eight of the stars in the constellation of Bootes are above the fourth magnitude, which means it is a great one to look for with the naked eye.

History of Arcturus

In 1635, Arcturus became the first star other than our sun to be observed during the day, when Jean-Baptiste Morin saw it through a telescope.

Vega

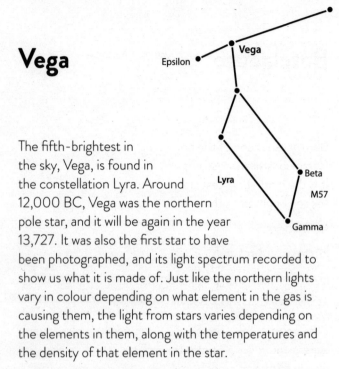

The fifth-brightest in the sky, Vega, is found in the constellation Lyra. Around 12,000 BC, Vega was the northern pole star, and it will be again in the year 13,727. It was also the first star to have been photographed, and its light spectrum recorded to show us what it is made of. Just like the northern lights vary in colour depending on what element in the gas is causing them, the light from stars varies depending on the elements in them, along with the temperatures and the density of that element in the star.

Vega is quite close, around twenty-five light years away from the sun, and is 2.1 times the mass of the sun.

HOW TO SEE IT

To find Vega in the northern hemisphere, look north-east during May, and if you see a bright bluish star that is probably Vega.

Betelgeuse

Betelgeuse is one of the bright stars that makes the Orion constellation, so it is also known as Alpha Orionis. It is the ninth-brightest star in the night sky, and the second brightest in the constellation of Orion.

HOW TO SEE IT

Betelgeuse is the left shoulder of Orion, so is easy to spot when it is in the sky. During January and February, it rises around sunset, making it a great star to look for in winter, in the northern hemisphere.

Betelgeuse shines with a red colour, because it is a red supergiant. It is 950 times the radius of the sun, and eleven to fifteen times as massive. This makes it one of

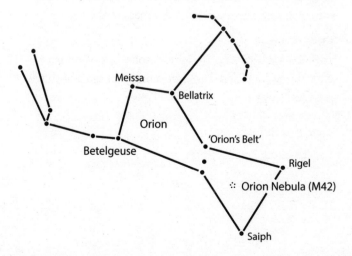

the biggest stars we know about. It is about eight to ten million years old, which is quite old for a supergiant, as we know they burn their fuel quickly and brightly. This means Betelgeuse might explode in a supernova, and it could happen any day now, according to some astronomers.

Estimates of the distance to Betelgeuse vary between 430 and 650 light years. This means it is entirely possible the star has already died, and we are yet to witness the explosion, as the light takes hundreds of years to reach us. When Betelgeuse's explosion finally reaches Earth, it will be incredibly bright in our sky, almost as bright as the full moon. This means it could be visible during the day for days or even weeks. While there are no guarantees it will happen within our lifetime, it is due to happen in the next thousand, or million years – the blink of an eye in astronomical terms. The explosion will not harm us on Earth, though, as the star is too far away for this to happen. It is estimated a supernova explosion has to be within fifty light years of Earth for us to be affected by it.

Because it is so bright and relatively nearby, astronomers have been able to picture the surface of a star using a set of telescopes in the Atacama Desert in Chile. It has also been observed in infrared, visible light and ultraviolet, and astronomers have found a giant bubble that boils on the surface of the star. It's also surrounded by a cloud of gas the size of our solar system.

Deneb

Deneb is an incredibly distant star, but it can still be seen in the night sky because of how bright it is in the first place, as it is a supergiant with a blue tint. It is 200 times the mass of the sun and its luminosity is somewhere between 50,000 and 200,000 times our own sun's. It is estimated to sit somewhere between 1.5 and 3.6 light years away.

HOW TO SEE IT

Deneb lies in the asterism the Summer Triangle, which is named this way because it is high in the sky during the summer evenings in the northern hemisphere. Finding Deneb is easy once you have located the other two stars in the asterism, Vega and Altair.

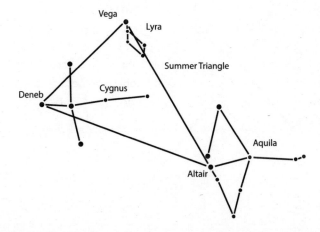

Albireo

Many of the single stars you see in the night sky are binary stars, but most of them, even on deeper inspection with binoculars, will look like a single point of light. Not Albireo, though. When seen through a powerful pair of binoculars, this binary system shows both the yellow and blue stars that make up the system – their names are Beta Cygni A and B. Through a telescope, this can be seen in even more stunning detail.

HOW TO SEE IT
Albireo is part of the Cygnus constellation, also known as the northern cross, and it is the fifth-brightest star in the constellation. It is the head of the swan, the furthest star from the centre of the constellation, so it is sometimes referred to as the beak star.

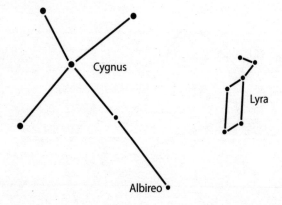

MAIN CONSTELLATIONS

Constellations are groups of stars that, taken together, form patterns in the sky, each named after an animal, god or a mythological or historical figure. The names of constellations differ for different cultures, because since people have been looking up, they have been naming patterns and telling stories about them, but the International Astronomical Union has named eighty-eight official constellations. The following constellations are a handful of these eighty-eight, in particular those that are best to look for with the naked eye from a city.

Orion

If you are reading this book, you can probably recognise Orion in the night sky. As previously mentioned, Orion is the most recognisable constellation during the dark nights of the winter sky in the northern hemisphere. Most people on Earth get a chance to see Orion in full at some point in the year, as it pops up over the horizon during summer in the southern hemisphere too.

HOW TO SEE IT

Orion is one of the first constellations you should learn to recognise, because it can be a useful one for star hopping to other constellations. It is also bright enough that it can be spotted in the middle of a city on a clear night, even if just the belt is visible.

Not only that, but if you grab a pair of binoculars you can reveal Orion is also home to some incredible nebulae, which can be a great first target for deep-sky observation, and even astrophotography.

The whole constellation of Orion is made up of fourteen stars in total. When the lines are joined between the stars, the shape looks like a hunter holding a bow and arrow, which is where the name came from, as Orion was a hunter in Greek mythology.

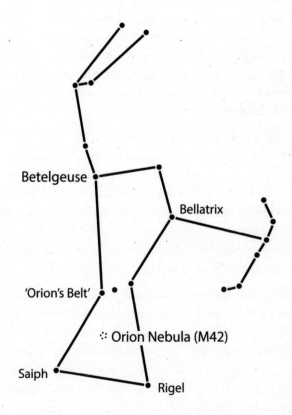

Three of the stars in Orion – Alnitak, Alnilam and Mintaka – make up the asterism Orion's belt, the most recognisable part of the constellation. Orion contains some of the brightest stars in the sky. The bright Betelgeuse, a red supergiant, makes the left shoulder, while the right foot is Rigel, another supergiant and the sixth-brightest star in the night sky. Unlike Betelgeuse, Rigel is a blue supergiant, but it is also likely to come to the end of its supergiant phase soon, either in a supernova or to become a white dwarf.

Orion is a great constellation to begin star hopping from. Going from the right of the belt to the left, carry on that line and you will reach the brightest star in the sky, Sirius, which is part of the constellation Canis Major. If you start at the right-hand star in the belt, Mintaka, then draw an imaginary line up to the top star in the bow and then continue this line, you will reach the star Aldebaran, which is in the Taurus constellation.

Going from Mintaka up to Betelgeuse, the left shoulder of Orion, continue this line and you will be able to find Castor and Pollux, the two stars that form part of the Gemini constellation. If you draw a line along Orion from Betelgeuse to Bellatrix, and continue on, this will point you in the direction of Cetus, the fourth-largest constellation in the sky.

Ursa Major

Many different civilisations came to the same conclusion, separately: that Ursa Major represents a bear. Ursa Major is the third-biggest constellation in the sky, and it can be seen for most of the year in the northern hemisphere. In the southern hemisphere, viewers will never glimpse the entire constellation, but the southern parts of the constellation can sometimes be seen.

Ursa Major is best known for a group of seven stars within it, which make the asterism known as the Big Dipper in the USA and the Plough in the UK. The shape of this

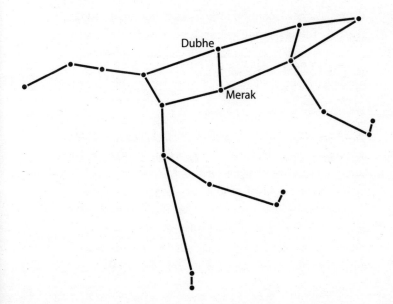

asterism looks like the Little Dipper, or Ursa Minor, which is itself a whole constellation. The Plough is always the first step when looking for Ursa Major in the sky.

HOW TO SEE IT
Ursa Major is found in the northern sky, so start by looking north. During summer in the northern hemisphere, it may appear close to the horizon, depending on what latitude you are at. During the winter, the handle of the Plough will look as though it is hanging down, while in spring and summer the bowl will be pointing downwards.

The four stars that make the bowl of the Plough include Dubhe and Merak, on the end furthest from the handle, which are pointer stars that you can use to locate Ursa Minor.

Once you have identified the Plough, you have found the tail of the bear and its back end. Extend the bowl forwards, away from the tail, and you will see five stars that make up the front of the bear. From there, you can see the bear's legs. In total, Ursa Major is made up of twenty-five stars.

Because Ursa Major lies away from the Milky Way's disk, it is a good place to spot other galaxies, but most of them are not visible with the naked eye. Messier 81, which is one of the brightest galaxies in the sky, lies next to Messier 82, above the bear's head. The Pinwheel galaxy is a spiral galaxy that can be found north-east of the constellation. Looking at these galaxies requires a small telescope, or powerful binoculars for Messier 81.

Ursa Minor

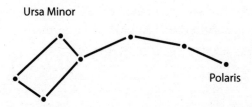

Ursa Minor

Polaris

The Little Bear constellation, or Ursa Minor, is a small version of the Great Bear, except it only contains seven main stars. It resembles a smaller version of the Big Dipper, which is why it is often called the Little Dipper. It can also be located using the pointer stars from the Big Dipper.

HOW TO SEE IT

Ursa Major is an important constellation because it contains the North Star, or Polaris. To find Polaris, you use the stars Merak and Dubhe, which are the far end of the Plough. Going from Merak to Dubhe, extend that line onwards for around 4.5 times the distance again, and you will reach Polaris.

Polaris is the end of the handle of Ursa Minor, so to find the rest of the constellation you must look left once you have reached Polaris, going from the pointer stars. The stars in Ursa Minor are relatively faint, compared to Ursa Major. It is also quite a small constellation. The two stars furthest from the pole, Kochab and Pherkad, are known as the guards or guardians of the pole star Polaris.

Ursa Minor is a circumpolar constellation in the northern hemisphere, and it cannot be seen from the southern hemisphere. It was first noted by the Greek astronomer Ptolemy, in the second century.

Ursa Minor is an exciting constellation in terms of the variety of stars involved. Polaris is a supergiant, and the brightest Cepheid variable in the sky. Beta Ursae Minoris, which is also called Kochab, is an old orange giant. Planets have been discovered orbiting around Kochab and three of the other stars in Ursa Minor.

Ursa Minor is home to Calvera, a neutron star that was discovered in 2007, and one of the hottest that astronomers have found. The temperature on the surface of Calvera is almost 200,000 degrees Celsius. It cannot be seen with the naked eye, so is not one of the seven bright stars you would see in the constellation, but it is a really exciting star. It is also the closest neutron star we have ever found, at between 250 and 1,000 light years away.

Cassiopeia

Cassiopeia lies on the other side of Polaris from Ursa Minor. It is a small constellation made up of five stars that form a letter W or M depending on which way up it is.

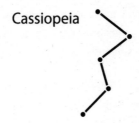

Cassiopeia

Polaris

HOW TO SEE IT

To find Cassiopeia, you must first know where Ursa Major and Ursa Minor are. Look inside the Plough, to the second star along in the handle, which is called Alioth, then find Polaris. Drawing an imaginary line from Alioth to Polaris, continue this line in the same direction around half the distance, and you will reach one of the middle stars of Cassiopeia.

Because of where it lies in the galaxy, in front of the disk of the Milky Way, on a dark night Cassiopeia can be surrounded by other stars which make it tricky to identify. This comes with practice.

Cassiopeia is named after a Greek queen of the same name, who was known for being vain. It is one of the forty-eight constellations that was first catalogued by Ptolemy in the second century, along with its neighbour Ursa Minor. Much like Ursa Minor, Cassiopeia is a circumpolar constellation in the northern hemisphere, which means it never sets. In the southern hemisphere, it is seasonal and can only be seen during summer.

The constellation is home to some of the most luminous stars we know of, including a yellow hypergiant that is 500,000 times brighter than the sun, of which only a dozen have been discovered in the Milky Way.

Cygnus

Cygnus, or the 'swan' constellation, is a bright constellation situated in the northern hemisphere and can be seen most of the year round, but is easiest to spot in summer when it is high in the sky. It contains an asterism called the Northern Cross, which is much larger than its counterpart, the Southern Cross. The Northern Cross is made of the brightest stars in Cygnus, so it is a great way to start when looking for the whole constellation.

HOW TO SEE IT

To find Cygnus, you must first look for the brightest star in the Northern Cross, Deneb. Deneb is one of the three stars in the northern hemisphere's asterism the Summer Triangle, alongside Vega and Altair. Knowing the Summer Triangle helps find the Northern Cross, so try and find this first (p. 105).

Then, looking from Altair to Vega, around halfway between the two, look towards the brightest star in that section of the sky. This star is Albireo, discussed earlier in this chapter (p. 126).

Albireo and Deneb make up the longest side of the Northern Cross, and between them is the central star. If you look to each side, you will find the two other stars that make up the cross. This is the backbone of the Cygnus constellation. Extending the arms of the cross

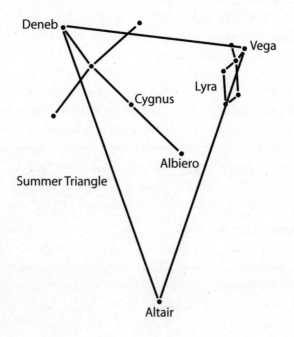

out in both directions, you will be able to find the rest of the Cygnus constellation. Deneb is the swan's tail and Albireo is the head, or the beak.

On a dark night, Cygnus can be a great way to find the Milky Way, as the plane of the galaxy's disk lies along the longest side of the Northern Cross. If the sky is dark enough, looking through binoculars will allow you to see fields and clusters of stars and nebulae that are littered around the Milky Way's disk near to Cygnus.

Cygnus is named after the swan that Zeus transformed himself into in order to seduce Queen Leda of Sparta, which led to her giving birth to Helen and Pollux, in Greek mythology. Pollux is the name of one of the stars in the Gemini constellation.

Canis Major

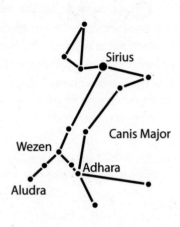

The brightest star in the night sky is Sirius in the constellation Canis Major, or the great dog. It is an easy one to find when it is visible, which tends to be whenever Orion is in the sky.

HOW TO SEE IT

To locate Sirius, you use the constellation of Orion, as the three stars in the belt of Orion point towards Sirius. After finding Sirius, Canis Major is easy to see, and it was included in Ptolemy's forty-eight constellations that he documented in the second century.

Canis Major is made up of around fifteen stars, with Sirius as the brightest, of course. Sirius marks the neck of the dog, with the head extending up from it. The second brightest in the constellation, called Adhara, marks the back legs. It is an extremely bright but distant star – the brightest source of extreme ultraviolet radiation in the night sky.

Canis Minor

Behind Canis Major follows Canis Minor, or the little dog. Together, the two constellations are said to be the dogs that loyally follow the hunter, Orion, through the night sky. Canis Minor is not a difficult constellation as it only consists of two major stars bright enough to be seen with the naked eye. Luckily, one of these is the eighth brightest in the night sky, Procyon.

To find Canis Minor, you need to find Orion and Canis Major. Roughly between Sirius and Betelgeuse, there are two stars which make up part of a constellation called Monoceros, or the unicorn. This is quite a faint constellation.

If you follow the line of Monoceros away from Orion, you will see another bright star: this is Procyon. To find the other half of Canis Minor, look a little towards Betelgeuse. Alternatively, if you can't find Monoceros, by looking for a triangle that consists of Betelgeuse, Sirius and another star, that points away from Orion and Canis Major, you will find Procyon.

Pegasus

The constellation of Pegasus is visible from July to January, but it is best to look for it in northern-hemisphere skies during autumn, which is spring in the southern hemisphere, from August to December. It is named after the winged horse Pegasus, from Greek mythology, who was said to have been born out of drops of blood on the ground.

HOW TO SEE IT
The Great Square of Pegasus is an asterism that lies mainly within the Pegasus constellation, and is made up of four bright stars. This makes it a great place to start when looking for Pegasus, the seventh largest of the constellations. The square is made up of the four stars Markab, Scheat and Algenib in the Pegasus constellation, and Alpheratz, which belongs to the constellation of Andromeda. From two of the points of the square extend Pegasus' head and front legs, pointing towards Cygnus.

To find the Great Square by star hopping, use the Plough or the Big Dipper to find the North Star, or Polaris. Then, draw an imaginary line from any star in the handle of the Plough through Polaris, to get to Cassiopeia. Continue this line through the Caph star, which is one to the four bright stars that make the Great Square of Pegasus.

Alpheratz is the brightest in the asterism and it can be seen towards the north-east. It also forms part of the Andromeda constellation.

Once you have found the Great Square of Pegasus, you can test the darkness of the sky by counting the stars held within the square. No stars mean your conditions are not great. If you see two, it means the magnitude you can see is 4.6, which means there is light pollution around. If you see eight stars, you can see stars with magnitude 5.5, and if you see thirteen stars you are seeing magnitude 6. The best you can expect to get with the naked eye would be a total of thirty-seven stars, which means you are seeing magnitude 6.5 – this is rare and would only happen in the darkest skies with no moonlight at all.

You can also use the Great Square to help you find your way to other interesting objects in the sky, like the Andromeda Galaxy by star-hopping from the star Alpheratz to two other stars in the constellation Andromeda, detailed in the following section. Andromeda is the nearest major galaxy to the Milky Way, and it is one of the most distant objects that can be seen with the naked eye.

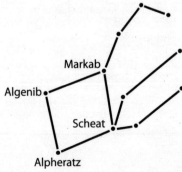

Andromeda

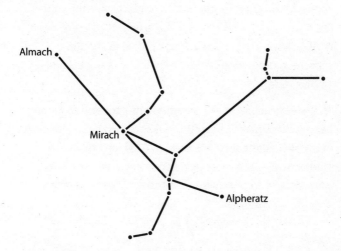

Not to be confused with the Andromeda Galaxy, the Andromeda constellation is named after the daughter of Cepheus and Cassiopeia in Greek mythology. She was said to be chained to a rock by her father as a sacrifice to sate a sea monster, then rescued by Perseus. In the sky, Andromeda lies beneath the constellations of Cassiopeia, Cepheus and Perseus.

HOW TO SEE IT
Finding the Andromeda constellation is easy once you have found Pegasus. From Pegasus, you can star hop to the Andromeda constellation, because the top-left star in the Great Square of Pegasus, Alpheratz, is the last star in Andromeda. This is also the brightest star in the Andromeda constellation.

Most of the stars are relatively low brightness, but it is not too difficult to spot because it sits away from the disk of the Milky Way, so the sky is not cluttered with other stars. The most interesting feature near the Andromeda constellation is the galaxy. It sits between the stars within the constellation, and between Cassiopeia and Andromeda.

The Andromeda Galaxy is named after the constellation, because it appears close to the Andromeda constellation in the sky. Also called Messier 31, M31, or NGC 224, Andromeda is the closest major galaxy, and the closest spiral galaxy, to our own – between 2.2 and 2.5 million

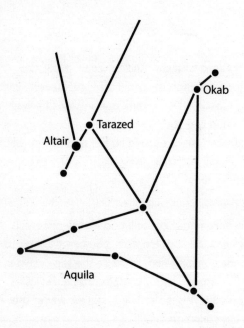

light years from Earth. Andromeda is the biggest galaxy in the local group, a name used to refer to those near the Milky Way, because it is made of about one trillion stars.

In the future it is predicted Andromeda and the Milky Way are going to collide and merge together, with our smaller galaxy being eaten up by our large neighbour. Don't worry, though: this is not going to happen for another four and a half billion years.

You can see the Andromeda Galaxy with the naked eye on a clear night with no moon, even in places with a little light pollution. It has an apparent magnitude of around 3.4. This makes it the most distant object that can be spotted with the naked eye.

Through binoculars, or through a telescope, you can spot the spiral arms of the Andromeda Galaxy. For this reason, it is one of the most popular deep-sky targets for astrophotographers.

Aquila

In Greek mythology, Aquila was an eagle that carried Zeus' thunderbolts. The constellation lies on the celestial equator, and its brightest star, Altair, makes one of the stars in the northern hemisphere's asterism the Summer Triangle.

HOW TO SEE IT

Altair is one of the closest stars to Earth, only seventeen light years away, making it the twelfth brightest in the night sky. Once you have found Altair, Aquila is made up of a group of stars arranged in a kite-like pattern, with Altair as the top of the kite, or the head of the eagle.

Aquila is also home to a star called Tarazed, or Gamma Aquilae. It is burning helium and is about to become a white dwarf, after burning all its hydrogen in just 100 million years. It has a magnitude of 2.7 and is 395 light years from the sun.

Aquila is home to some interesting deep-sky objects including planetary nebulae, called NGC 6804, NCG 6781 and NCG 6751. These can be an exciting target for people observing with a telescope.

In about four million years' time, the NASA probe Pioneer 11 is expected to pass close to the star Lambda Aquilae, the star opposite Altair in the constellation. Pioneer 11 passed Jupiter and Saturn in the 1970s, left the solar system in 1990 and lost communication with NASA in 1995.

Lyra

Once you've found the Summer Triangle, Lyra is the third constellation you can spot. Its name comes from a lyre, an instrument in Greek mythology that belonged to Arion or Odysseus. In Arabic astronomy, Lyra, Aquila and Cygnus all represented three birds.

Vega, which is the brightest of the three stars in the Summer Triangle, is a circumpolar star at greater latitudes than 52 degrees north. The constellation of Lyra can be seen in the northern hemisphere for most of the year, high in the sky from spring until autumn. In the southern hemisphere, it appears low in the sky during winter.

Another star in Lyra, Beta Lyrae, is actually a multiple-star system made up of a triple-star system, Beta Lyrae A, and two single stars, B and C.

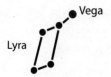

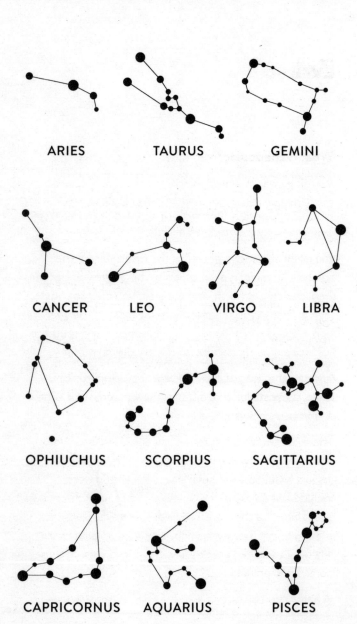

ARIES

TAURUS

GEMINI

CANCER

LEO

VIRGO

LIBRA

OPHIUCHUS

SCORPIUS

SAGITTARIUS

CAPRICORNUS

AQUARIUS

PISCES

Zodiac

What is the zodiac?

Astrology was invented thousands of years ago, and many people do not know where it actually comes from. Remember the ecliptic? It is the line the sun traces in the sky, which is followed, not exactly but closely enough, by the planets too.

Well, there are a set of constellations that pass through the ecliptic, too, which are the zodiacal constellations. As the night sky changes over the year, the constellations in each part of the sky change. Star signs come from the constellation that the sun passes through, during specific times of the year. The idea is that, if the sun was in a specific constellation the day you were born, that is your star sign.

There are twelve famous zodiacal constellations. There are more than twelve constellations on the ecliptic, it's just that some are smaller and fainter, so were not included when astrology was invented thousands of years ago. There is a thirteenth zodiacal constellation that is big and bright, however, called Ophiuchus, but it was left out by astrologers because the number thirteen was considered unlucky.

Whether you believe in astrology or not, most people

know their own star sign. However, the star sign you think you are could be wrong. This is because when the star signs were assigned to specific dates, it was thousands of years ago, and the stars have moved in the sky since then. Most people still use these outdated cut-offs, but the Earth's position in space has moved in the intervening thousands of years, so the star signs have changed, too. For example, if you were born in the middle of February in ancient Greece, you would have been an Aquarius, but today you would be a Capricorn.

This is not an astrology book, so you will not be able to learn anything about the suggestion that the position of Earth on your day of birth has any effect on your life, your relationships or your personality. What you will learn, however, is to identify each of the zodiacal constellations in the sky, and what kind of stars they have in them.

You might expect the zodiacal constellations would be visible during roughly the time when those with that star sign were born, but this is wrong. Remember, the star signs are in the plane of the axis of rotation of the Earth around the sun. They are also assigned when the sun is in that constellation, which means they are in the same part of the sky as the sun, so would only appear in the day when the light from the sun will outshine them. They are best seen six months later, when that part of the sky is dark.

Before learning about the zodiacal constellations, it is important to understand why their positions are changing in the sky. Earlier in the book (p. 80), we talked about how the axis of rotation is slowly shifting, in a process called precession. Imagine the Earth is like a spinning top, and it traces out a cone shape in its axis of rotation as it spins around. This process takes around 25,700 years. This explains why Polaris will not always be our North Star. The celestial North Pole is moving with respect to the galaxy around us, so the positions of the stars in the sky appear to move. But this is not why the zodiac constellations are moving.

Remember, the zodiac constellations lie in the plane of orbit of Earth around the sun, the ecliptic. Imagine we are at a theme park, on a train that takes us around in a huge circle, with the sun in the middle. On the outside are all the constellations of the zodiac. We cannot see them when they are on the same side of Earth as the sun, because the sun's light will block out our view of them. This means the zodiac constellations we can see at any point do not depend on the axis of rotation of Earth, but the position of Earth in its orbit around the sun.

To understand how the constellations move, we need to understand the way a year is defined. If one year started when the Earth was at one point in its orbit, and ended when it returned to that exact point, then the seasons would slowly change over thousands of years. Instead, we define a calendar year by the degree to which the Earth is tilted towards the sun, which does roughly, but not perfectly, match up with the time it takes to complete one full orbit of the sun. This is because of precession.

We know the tilt of the Earth towards the sun by the ratio of day to night at each point on Earth. The equinoxes happen when the day and night are equal all over Earth, while the northern hemisphere summer and winter solstices occur when the northern hemisphere is tilted fully towards or fully away from the sun. A tropical year is measured between two consecutive spring equinoxes.

A sidereal year, which is the time it takes to complete one full orbit of the sun, is actually twenty minutes longer than a tropical year. This means on the same date from one year to the next, the Earth will be in a slightly different position in its orbit.

In our lifetime, precession has no impact on the way the sky looks. But, over thousands of years, it shifts the perspective of what we can see when. For example, in January this year the sun was in Capricornus, but in around 13,000 years' time, when Earth is halfway through a cycle of precession, the northern hemisphere will be tilted away from the sun when the Earth is at the opposite side to the sun in its orbit. This means Capricornus will be in full view in the night sky.

This is why, when astrology was invented thousands of years ago, the position of the constellations in relation to the sun during specific months was different from what we see now.

Capricornus

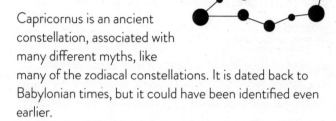

Capricornus is an ancient
constellation, associated with
many different myths, like
many of the zodiacal constellations. It is dated back to
Babylonian times, but it could have been identified even
earlier.

Capricornus is close in the sky to other constellations
associated with water, like Aquarius and Pisces, and the
name comes from the Latin for 'horned goat'. It is known
as a sea goat and has been seen as a hybrid of a goat and
a fish since the Middle Bronze Age. The constellation
is linked to the Greek god Pan, who escaped a monster
called Typhon by diving into a river and trying to turn
himself into a fish so he could get away. Because he was
rushing and so scared of the monster, he was only able to
turn himself half into a fish, and the top half remained a
goat.

Looking for Capricornus, you might not see a sea goat
straight away. The pattern the stars make resembles
more of a squashed triangle, but as ever with astronomy,
it helps to be imaginative.

WHEN TO SEE IT
Capricornus can be seen between July and October,
but is highest in the sky in early September. It is a faint

constellation, with a magnitude of around 3, so it is not easily spotted in busy cities with a lot of light pollution. If you are keen to spot Capricornus, take some time to drive away from light pollution one clear evening, or get yourself as high above lights as possible.

HOW TO SEE IT

To find Capricornus, start by finding the Summer Triangle asterism. Once you have spotted it, draw a line from Vega through Altair, and keep going in that direction until you see an arrow-shaped constellation in the southern sky. It will be highest in the sky around 10 p.m. local time, in early September. If you are spotting it from the southern hemisphere, it appears upside down compared to the way it looks from the northern hemisphere, and it will appear in the northern sky.

If you can't find the Summer Triangle, you can star hop to Capricornus by finding the constellation Cygnus, or the Northern Cross asterism. Start with the bright star Deneb in Cygnus, then go to Epsilon Cygni, and keep going in this direction to find Capricornus.

The Tropic of Capricorn refers to the circle of latitude around Earth that marks the southernmost point where the sun can be seen directly overhead, which lies 23.5 degrees south. The number 23.5 comes from the axis of rotation of the Earth, compared to the orbit of Earth around the sun.

The Tropic of Capricorn is also the line where on 21 December, the winter solstice in the northern hemisphere, the sun is directly overhead. The line is named after Capricornus because over 2,000 years ago, the sun was in Capricornus at that time. Now the sun is in Sagittarius at this time, but it remains the Tropic of Capricorn because of the historical association.

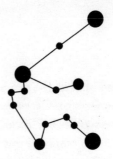

Aquarius

The name 'Aquarius' comes from Latin, meaning 'water-carrier', and it is said to represent a water carrier pouring water from the mouth of the jar. From the Middle East, Aquarius appears in the east just before the start of the rainy season, which is considered a possible reason why the constellation is associated with water.

Other than Capricornus, water-themed constellations near Aquarius include Cetus, the whale; Pisces, the fish; and Eridanus, the river. For this reason, this part of the sky is called the sea. The brightest star in the sea region of the sky is called Fomalhaut, and it's in the constellation of Piscis Austrinus, the southern fish.

In Greek mythology, the constellation Aquarius is linked to Deucalion, Prometheus' son, who expected a flood and built a ship for his wife. It is also associated with Ganymede, a cup-bearer for the gods.

WHEN TO SEE IT
Visible around the same time as Capricornus, Aquarius is easiest to spot during autumn in the northern hemisphere, which is spring in the southern hemisphere. It is high in the northern sky in the southern hemisphere, and in the southern sky in northern latitudes.

HOW TO SEE IT

To star hop to Aquarius, it is easiest to first spot Capricornus using the star-hopping techniques mentioned previously. Then look to the north-east of Capricornus. The immediate stars in this region are part of Aquarius. In other words, if you consider the outline of Capricornus as kind of a triangle, Aquarius lies next to the longest side of the triangle, as if the water bearer is leaning on Capricornus.

You will need a dark sky to spot Aquarius, so city dwellers will have to travel to tick this one off their list. There are no particularly bright stars in Aquarius, but it is home to some interesting ones. Alpha Aquarii and Beta Aquarii are both yellow supergiants, extremely bright and large, but they appear dim in our sky because they are so far away.

If the sky is dark enough, you can also spot an asterism inside the constellation of Aquarius: the water jar, which is sometimes called the Y of Aquarius. It is made up of thirty faint stars that form a zigzag pattern

Astronomers have spotted exoplanets orbiting some of the stars in Aquarius, including Gliese 876, which is one of Earth's closest stars and has four planets orbiting it. The star, which is invisible to the naked eye, is a red dwarf and it's the third-closest star with known planets orbiting it.

Pisces

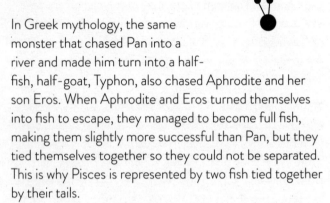

In Greek mythology, the same
monster that chased Pan into a
river and made him turn into a half-
fish, half-goat, Typhon, also chased Aphrodite and her
son Eros. When Aphrodite and Eros turned themselves
into fish to escape, they managed to become full fish,
making them slightly more successful than Pan, but they
tied themselves together so they could not be separated.
This is why Pisces is represented by two fish tied together
by their tails.

WHEN TO SEE IT

Pisces is easiest to spot during November to early
December, when it reaches its highest point in the sky.

HOW TO SEE IT

Unsurprisingly, Pisces also lies in the sea region of the
sky. As with the other constellations in this busy watery
region, you need a dark sky to spot Pisces. If you have
found Aquarius and Cetus, Pisces is found to the north-
east of Aquarius and north-west of Cetus.

Another way to find Pisces is using the Great Square
of Pegasus, an asterism we discussed earlier in the book
(p. 143). If you find the square, then look south or away
from Andromeda: you will be able to find an asterism
called the Circlet of Pisces, which is one of the fish's

heads. From there, look away from the square of Pegasus and you will see the rest of the fish: this is the western fish. If the western fish is roughly parallel to the bottom of the square, then the eastern fish is parallel to the left side. It points upwards towards Andromeda.

Together, the two fish look like they make a 'V' shape in the sky. The star at the centre of the V is called Al Risha, and although it is not one of the brighter stars in the sky, or even in the constellation, it is easy to spot because of its position. Al Risha, also called Alpha Piscium, is a binary star system, located about 311 light years from Earth.

On 21 March, or the vernal or spring equinox in the northern hemisphere, the sun appears in front of the constellation of Pisces. A tropical year is often defined as the 365 days between a March equinox and the next one, so Pisces is often considered the start of the zodiacal constellations. Even though the spring equinox is currently when the sun passes Pisces, it is still called the First Point of Aries, because thousands of years ago the sun was in Aries during the spring equinox. When this next changes, in around 600 years, Aquarius will become the constellation in which the sun sits during the spring equinox. In astrology, some say this is when the Age of Aquarius starts.

As well as the two eastern and western fish that make up Pisces, there is another constellation called Piscis Austrinus, or the south fish.

Aries

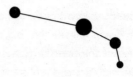

In Greek mythology, Jason and the Argonauts went on a mission to find a golden fleece made from a ram with golden wool and wings. This ram is represented by the constellation of Aries.

WHEN TO SEE IT
Aries pops up in the sky a few weeks after Pisces, making it visible throughout late November and December, when the constellation can be found during most of the night.

HOW TO SEE IT
To star hop to Aries, first you need to find the North Star or Polaris, and the constellation of Cassiopeia. Draw an imaginary line from Polaris through the star Segin, which is found on one of the ends of Cassiopeia, next to the star Ruchbah. Continue this line onwards and it will point you in the direction of Aries.

Again, like many of the zodiacal constellations, Aries is not particularly bright, so it is best to search for the ram when there is no moonlight and as little light pollution as possible. Aries' three main stars, Hamal, Sheratan and Mesarthim, make up the ram's bust. The ram's head can be found halfway between the Square of Pegasus and the Pleiades star cluster.

The First Point of Aries actually lies in the constellation of Pisces these days, but it has kept its old name. It is one of two points on the celestial sphere that intersects with the ecliptic, and it marks zero degrees right ascension and declination. The sun is found there during the March equinox.

If you are interested in deep-sky objects, Aries is not the constellation for you. There is little to be seen with a telescope within Aries, other than the binary system of Mesarthim. Also called Gamma Arietis, this double star system was discovered in 1664 by Robert Hooke, and the two stars are far enough apart that they can be seen as two stars using a small telescope. It takes over 5,000 years for the two stars to orbit each other.

The closest star to Earth within Aries cannot be seen with the naked eye. You need a big telescope to spot the brown dwarf called Teegarden's star, even though it is the twenty-fourth-closest star to Earth. It is about twelve light years from Earth and was discovered in 2013 through a set of data that had been first gathered to track near-Earth asteroids.

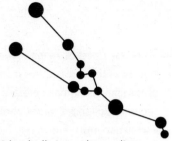

Taurus

Taurus has been associated with a bull since the earliest astronomers, and the constellation has been recognised as as far back as the Early Bronze Age. In Greek mythology, it is associated with the bull that Zeus turned himself into when he abducted the princess Europa.

WHEN TO SEE IT
Taurus is visible from November to March, but the best time to look for it is January. Because of its relatively bright stars, it is one of the zodiacal constellations that can be easily spotted, even in urban areas.

It is best known for the red giant that sits inside it – a star called Aldebaran. Aldebaran is sixty-five light years away from the sun and is the fourteenth-brightest in the night sky. It's the brightest star in Taurus, and a good place to start when looking for the constellation.

HOW TO SEE IT
Taurus and Orion are constellation neighbours. You can find the red glow of Aldebaran by first finding Orion. Start with the bottom-right star in Orion, Rigel, then draw a line up through Orion's bow, and you will see a bright, red star, which is Aldebaran. Alternatively, if you know which direction south-west is from Orion, look that way and you will also find the star.

Another interesting star within Taurus is Elnath, which comes from an Arabic word meaning the 'butting', or 'bull's horns'. Elnath is 700 times brighter than the sun and lies 130 light years away.

Taurus lies along the plane of the Milky Way, so the sky near by is rich with deep-sky objects including nebulae and clusters of stars. In fact, Aldebaran falls inside an asterism called the Hyades, a V-shaped pattern that is sometimes referred to as the bull's face. From the face, you can search for the remaining stars in the constellation of Taurus. The Hyades is a cluster of stars, so looking through a telescope can reveal a group of around 100 stars that is over 600 million years old. Unlike many stars that appear next to each other from Earth but are in fact incredibly far apart, the Hyades cluster is made up of stars that were all formed at the same time, from the same stuff.

Another cluster near the Taurus constellation is the Pleiades. The Pleiades is a fascinating deep-sky target to look for and one of the easiest to spot in the sky. More detail on the Pleiades cluster can be found later in the book (see p. 194).

During autumn, if you look towards the Taurus constellation, you can find the Taurids meteor shower. Normally, it peaks in late October or early November, but this varies. The Taurids meteor shower comes from two separate showers, with a southern and northern part. The northern Taurids come from an asteroid, while the southern Taurids come from a comet, called Encke.

The theory is that the comet Encke was part of a much bigger comet that fell into smaller pieces between 20,000 and 30,000 years ago. Because they often fall at the end of October, the meteor shower is also called Hallowe'en fireballs.

Gemini

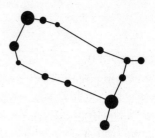

Gemini can be distinguished by its two brightest stars, Castor and Pollux. The 'twins', Castor and Pollux, that make up part of the constellation of Gemini refer to a pair in Greek mythology who were in fact only half-brothers. Although the word Gemini means 'twins' in Latin, Castor was the son of King Tyndareus and Pollux was the son of Zeus. Both shared the same mother, Queen Leda of Sparta. In Babylonian astronomy, the stars Castor and Pollux were known as the Great Twins.

As with twins in real life, it can be tricky to tell Castor and Pollux apart. Castor is further north, while Pollux is slightly lower in the sky, but is brighter. Gemini is made up of many pairs of binary stars, which is fitting, given its theme of twins. However, most of the binary stars still appear as one source of light through small telescopes, so it can be difficult for amateur astronomers to see them.

WHEN TO SEE IT

Gemini appears in the sky in mid-August, but the best time to see it is during January and February, when it stays in the sky for most of the night. By April and May, it can only be seen just after sunset, in the west of the sky, where it sets soon after the sun. During June and July, the sun is in Gemini, so it cannot be seen.

HOW TO SEE IT

The best way to spot Gemini is to look for its famous twins. You can do this by using the Plough or Big Dipper. Draw a line from the handle through the bowl of the Big Dipper, from Megrez to Merak. Keep this going and you will see Castor and Pollux.

As always, there are many ways to star hop to Gemini. Another way to find the twins is by looking from the bright star Rigel, Orion's right foot, to Betelgeuse, in Orion's shoulder. Drawing a line from Rigel to Betelgeuse, continue this until you see two stars close together. This is Castor and Pollux. These make the heads of the twins, and the rest of the constellation lies closer to Orion. As one of the brightest constellations in the zodiac, there are a lot of chances to see Gemini even in slightly light-polluted areas. You do not have to venture to the darkest skies in the country to see Gemini, as long as you have a clear night.

The moon also passes through Gemini a few days every month, just like it passes through every constellation in the zodiac. The moon very roughly follows the ecliptic in the sky, not as closely as the planets, but within plus or minus five degrees, which is around ten moon diameters in the sky. If you consult a chart beforehand, you can find when the moon will be passing through Gemini, and sometimes it can be seen between Castor and Pollux, making it easy to find them.

We wouldn't expert Gemini to have many deep-sky objects because, unlike Taurus, it isn't on the plane of the

Milky Way. Still, it has a few which include the Medusa Nebula, a neutron star called Geminga, and the Eskimo or Clown Face Nebula. During December, the Geminids meteor shower appears in the constellation of Gemini.

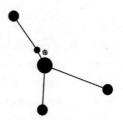

Cancer

Although Cancer is quite a faint constellation, it has been considered important for thousands of years because of its position. The sun appeared in Cancer during the summer solstice, which made it important to those in the northern hemisphere.

In Greek mythology, Cancer is linked to the crab that popped up when Hercules fought a water monster called the Hydra. The crab bit Hercules, so he crushed it, angering the goddess Hera, who put the crab into the stars because she hated Hercules. But Cancer has not always been linked to crabs. The Egyptians said it was a scarab, and others have described it as a crayfish or a lobster.

Like the Tropic of Capricorn, the Tropic of Cancer is named after the line on the Earth at which the sun appears directly overhead on the summer solstice in the northern hemisphere, usually on 21 June. This is when the Earth's northern hemisphere is tilted to its greatest degree towards the sun, giving the northern hemisphere the longest day of the year. Of course, now the sun is in the border of Gemini and Taurus on the summer solstice, but the name is kept anyway.

As we said, Cancer is faint. It is the faintest part of the zodiac, which means seeing it in light-polluted cities could be difficult. The brightest star in Cancer is magnitude 4, so it is advisable to go to an area with less light pollution to spot the constellation. A country sky on a clear night is great. In the southern hemisphere, Cancer appears in the north, while in the northern hemisphere it appears due south in the sky.

WHEN TO SEE IT
The best time of year to spot Cancer is March, April and May.

HOW TO SEE IT
To star hop to Cancer, first identify Castor and Pollux in Gemini, then the bright star Regulus, which lies in Leo. For how to star hop to Regulus, check the Leo section, on the next page. The constellation of Cancer lies directly between Gemini and Leo. If your sky is dark enough and there is no moonlight, it is easy to find Cancer.

While Cancer itself is faint, once you've found it you can see one of the brightest star clusters in the sky, the Beehive Cluster, which is also known as Praesepe. If you grab a pair of binoculars and look at the Beehive Cluster, which looks a little like a cloud to the naked eye, you will be able to see the 200 stars that make up the cluster. The best time to see the Beehive Cluster is February.

Leo

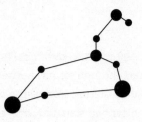

Leo, which is Latin for 'lion', represents the lion in Greek mythology that was killed by Hercules, in the first of his twelve labours. The constellation has been associated with lions since Sumerian and Babylonian times, when the sun was in Leo during the solstice. Now the sun is in Leo during the end of August and beginning of September.

WHEN TO SEE IT

The best time to see Leo is around the end of March, April and May. It is up most of the night during these months, making it a great spring constellation in the northern hemisphere. The best way to find Leo is by first looking for an asterism within it, called the Sickle, which looks like a backwards question mark. The bottom of this question mark is the star Regulus, which can also help find Cancer.

HOW TO SEE IT

To find the Sickle, you can start at the Plough or the Big Dipper. If you remember how to hop to Polaris, using the pointer stars (see p. 107), Leo is just looking in the other direction. Instead of going from Merak to Dubhe, which takes you to the North Star, go from Dubhe to Merak and keep that line going to find the Sickle. The Sickle makes up the head of the lion of Leo, and the rest of the constellation extends out from Regulus, away from Cancer and Taurus, towards Gemini.

The star at the lion's neck, Gamma Leonis, is a binary star. Through a small telescope, it can be possible to see this as two separate light sources. The best time to look for the two stars of Gamma Leonis is when the stars do not appear twinkly in the sky, because it means there is a lower amount of atmospheric turbulence.

One of the most exciting meteor showers, the Leonids, appear in Leo during the middle of November. However, it is not the most dependable. The Geminids are much more reliable for a good display of meteors, although there are occasions when the Leonids put on a huge display.

For those keen on deep-sky objects, Leo is home to a pair of galaxies called M65 and M66, which can be seen with a small telescope. Slightly trickier to find are M95 and M96, another galaxy pair in Leo.

Virgo

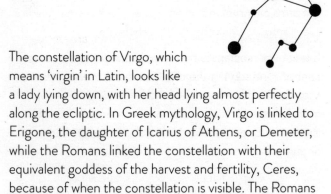

The constellation of Virgo, which means 'virgin' in Latin, looks like a lady lying down, with her head lying almost perfectly along the ecliptic. In Greek mythology, Virgo is linked to Erigone, the daughter of Icarius of Athens, or Demeter, while the Romans linked the constellation with their equivalent goddess of the harvest and fertility, Ceres, because of when the constellation is visible. The Romans also associated Virgo with the goddess of justice, Astraea.

During the autumn equinox, the sun lies in Virgo. It used to occur in Libra, which is why the equinox is referred to as the First Point of Libra. In the year 2440, the autumn equinox will lie in Leo.

WHEN TO SEE IT
Virgo is best spotted during late May, June and July. As the biggest constellation in the zodiac, and the second biggest in the night sky after Hydra, it can be difficult to find the pattern made from all the stars in Virgo. But finding where to start is easy.

HOW TO SEE IT
Virgo's brightest star, Spica, lies almost perfectly along the ecliptic. To find Spica, you can star hop from Ursa

Major and Bootes. Follow the curve of the Plough or Big Dipper within Ursa Major, which points you in the direction of the star Arcturus within Bootes. Follow this curve and you will be able to find Spica, which is one of the arms of Virgo.

Virgo is home to a lot of exoplanets. There are thirty-five known exoplanets that have been spotted orbiting twenty-nine stars in Virgo. In particular, the star 70 Virginis was home to one of the first exoplanets ever discovered, a planet seven and a half times the size of Jupiter. But home to one of the biggest exoplanets ever detected is the star Chi Virginis, which has a planet eleven times as massive as Jupiter orbiting it. Another star in Virgo, 61 Virginis, which looks a lot like our sun, is home to three known planets – two are the mass of Neptune and one is a super-Earth, which means it is more massive than Earth, but not quite as big as the ice or gas giants in our solar system.

Virgo is home to a huge galaxy cluster, meaning it is home to a lot of galaxies. The Virgo Cluster lies at the centre of what is called the Local Supercluster, which is thought to contain 10,000 galaxies. For most amateur astronomers, the majority of these galaxies are difficult to detect unless you have the biggest telescopes.

Libra

Until Roman times, Libra was not
considered its own constellation at all.
Instead, it was part of Scorpius, and was
thought to show the scorpion's claws. Now, of course,
Libra represents scales or balance. The Romans believed
the scales were being held by the goddess Astraea,
represented by the nearby constellation Virgo, and
they are linked to scales of justice. It is the only zodiac
constellation that does not represent a living thing.

WHEN TO SEE IT

The change from claws to scales is not fully understood.
It could be the case that the balance represents the
autumn equinox, when the day and night are equal,
because the sun enters Libra during that time, three
thousand years ago. Now the sun passes in front of
Libra from the end of October to the end of November.
The best time to see Libra is during the summer in the
northern hemisphere – winter in the southern – when it
appears in the sky during most of the evening.

Although Libra is quite faint, there are two bright stars
that point to its position in the sky, Spica and Antares.
Spica, in the constellation of Virgo, and Antares, in
Scorpius, surround the scales of Libra.

HOW TO SEE IT

To star hop to Libra, find the two stars just mentioned and search in the middle for a pair of fainter stars. These make up the two ends of the scales, and they have some of the best names out of any of the stars in the sky: Zubenelgenubi and Zubeneschamali.

Zubenelgenubi is the second brightest of the stars in Libra, and it is a binary system that can be seen as two stars when you look through binoculars. It is 77 light years from Earth and has a blue white glow. Its name means the southern claw in Arabic, from when the stars were considered part of Scorpius.

Zubeneschamali means the northern claw, and it is the brightest in the constellation. The star is 185 light years from Earth, sitting almost exactly on the ecliptic, and it is a slightly green colour.

Libra is also home to the oldest known star in the universe, Methuselah or HD 140283, which is thought to be almost as old as the universe itself. It is a red subgiant star, which means it is brighter than normal stars but not quite as bright and large as giant stars – but it is getting there. The star is thought to be about 190 light years from Earth, and it appears to be older than the universe. This is because researchers believe it is 14.5 billion years old, while the universe is thought to be 13.8 billion years old. Of course, it is not older than the universe, and there are uncertainties in both the numbers.

Three of the stars in Libra have known exoplanets. Gilese 581 is known to have at least three planets, but possibly six, orbiting it. Some of these may lie in the star's habitable zone, a distance at which liquid water could exist on the planet's surface, making it a potential home to life as we know it on Earth.

Scorpius

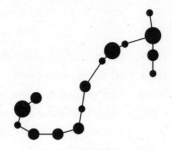

In mythology, Scorpius represents the scorpion that stung and killed Orion the hunter, and the two constellations are found on opposite sides of the night sky, so they can never be seen on the same night.

One story says Orion had boasted that he would kill every animal on Earth, but the goddess Artemis wanted to protect all the creatures, so she sent a scorpion to kill him.

WHEN TO SEE IT
The sun only spends about a week in the constellation of Scorpius, at the end of November. The J-shape of Scorpius sits next to Libra in the night sky, and it is easiest to spot by first finding the star Antares. Scorpius is best to spot in July and August, when it appears during most of the evening.

HOW TO SEE IT
It is easy to find Antares, one of the twenty brightest stars in the sky, if you know where to look. Star hopping, as usual, can be done in many different ways, but one method is to first find Altair in the Summer Triangle and Arcturus in Bootes, then look halfway between them and down until you see a bright, red star.

The appearance of the star Antares in the night sky is associated with summer in the northern hemisphere. Its name means rival of Mars, and it is a bright, red supergiant around 9,000 times as bright as our sun. The best time to see Antares is at the end of May, when it is in opposition to the sun.

Once you have spotted Antares, which makes the scorpion's head, the rest of the constellation is easy to spot because it makes a distinctive J shape. The two stars that make the stinger, Shaula and Lesath, lie in front of the Milky Way.

Because of its position relative to the Milky Way, there are lots of deep-sky objects to spot within and near Scorpius. These include clusters like the Butterfly Cluster and the Ptolemy Cluster, among other globular clusters. Both of these, also known as M6 and M7, are visible to the naked eye as fuzzy patches, but through binoculars they appear as groups of individual stars. There are also nebulae near Scorpius, like the Bug Nebula and the Cat's Paw Nebula – both star-forming regions. This means Scorpius is a great target for astrophotographers, or those with telescopes looking to spot something further away.

Ophiuchus

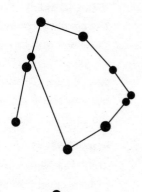

The chances of someone thinking their star sign is Ophiuchus are very low. Ophiuchus, the serpent bearer, is one of the thirteen constellations of the zodiac, but it is not included in most of the lists of zodiac signs in astrology. Recently, some people have been confused by articles suggesting Ophiuchus has been added to astrology, but that depends on what kind of astrology you believe in, if any at all.

Nevertheless, Ophiuchus is and has always been one of the thirteen zodiacal constellations, even if nobody has ever heard of it. People who are born between the end of November and the middle of December are born when the sun is in Ophiuchus. In mythology, Ophiuchus is related to the god of healing, Aesculapius. He was thought to carry a snake around to bring him wisdom and knowledge. The name Ophiuchus comes from two Greek words meaning 'serpent' and 'holding'. The snake is represented by the nearby constellation of Serpens.

WHEN TO SEE IT

The constellation is quite large, so the sun spends around three weeks in Ophiuchus, compared to the week it spends in nearby Scorpius, for example. The best time of year to view Ophiuchus is during summer in the northern hemisphere, especially in June and July, which is winter in the southern hemisphere.

HOW TO SEE IT

Star hopping to Ophiuchus is easy when you have found Antares: just look along the ecliptic and to the north slightly, to find the large constellation. It is also south of the constellation Hercules.

The brightest star in Ophiuchus is called Alpha Ophiuchi, often called the head of the charmer, with a magnitude of around 2. It is a binary star system that takes just under nine years for the two stars to orbit each other, and it is thought the stars are about 2.4 and 0.85 times the mass of the sun. The system is about forty-nine light years from Earth. One of the two stars is rapidly rotating, almost fast enough to cause it to break apart.

Another star, RS Ophiuchi, is a star that is getting brighter and dimmer at regular intervals; something called a recurrent nova. It is thought the star could become a supernova soon.

Like its neighbouring Scorpius, Ophiuchus is near the band of the Milky Way, so it is a great place to look for deep-sky objects, especially through binoculars on a night when there is no moonlight. Ophiuchus is home to the stunning star cloud Rho Ophiuchi, a multiple star system that is one of the few star clouds visible through binoculars. If viewed through binoculars, you can see a magnitude 5 and two companion stars.

Sagittarius

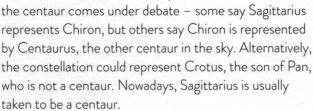

In Greek mythology, Sagittarius is a centaur, meaning he is half-human and half-horse. But the identity of the centaur comes under debate – some say Sagittarius represents Chiron, but others say Chiron is represented by Centaurus, the other centaur in the sky. Alternatively, the constellation could represent Crotus, the son of Pan, who is not a centaur. Nowadays, Sagittarius is usually taken to be a centaur.

WHEN TO SEE IT

Because it is quite a faint constellation, you will need a dark sky to see Sagittarius, so city-dwellers will likely have to travel to spot this one. It's best seen in August or September, next to the band of the Milky Way that stretches across the sky. A bulge in the middle marks the centre of the Milky Way, which is where Sagittarius can be found.

Another way to find Sagittarius is using the Summer Triangle asterism. Draw an imaginary line from Deneb through Altair, and keep it going to point towards Sagittarius.

It can be difficult to spot the shape of the centaur in Sagittarius, but an asterism called the Teapot can be easier to find. The Teapot is in the western half of

Sagittarius, and learning to identify its shape can help in star hopping to lots of other constellations.

Sagittarius contains the centre of the galaxy, which means it is home to a huge array of deep-sky objects. There are lots of nebulae and globular clusters, but there are also lots of stars too, which makes it tricky to see the clusters.

4

THE MILKY WAY

Every star you see in the night sky belongs to the Milky Way, so whenever you look up, whether you are seeing the moon, the sun or bright stars, you're seeing part of our galaxy. The only exception is the Andromeda Galaxy, which we will talk about later (see p. 198).

Our solar system lies on the edge of the Milky Way, which is a huge spiral shape with a bulge in the middle and arms coming out. We lie on one of the spirals, called the Local Arm or the Orion Arm. This means that if you look in a certain direction from Earth, you can see a vast collection of stars in the centre of the galaxy. The other directions are filled with fewer stars, but still have interesting things to look at.

The Milky Way contains at least 100 billion stars and is 100,000 light years wide. We sit around 26,100 light years from its centre.

When astronomers talk about seeing the Milky Way, they mean the spiral arm close to the central part of the galaxy. From Earth, it can be seen easily with the naked eye, but not in places with light pollution. In fact, an atlas of the night sky published in 2016 said that 80 per cent of the world's land areas are too light-polluted to see the Milky Way.

This means the Milky Way should be a treat you save for yourself when you are going on holiday somewhere with hardly any light pollution. For example, in New Zealand, the Mount Cook observatory is an amazing place from which to see the Milky Way, and is in a light-protected area. From the southern hemisphere, you can see more detail in the structure of the Milky Way, including the flattened bulge surrounded by the collection of stars that spiral out from the centre.

Anywhere with as little light pollution as possible works best. Use an online dark-sky finder to plan the places you can get to with as little light pollution as possible. In the UK, the only places with no light pollution are in the north of Scotland. However, if you find a place with low light pollution compared to the city you live in, the chances are that you will still get a good view of the Milky Way.

To find it in the sky should be easy if it is dark enough. But the Great Rift of the Milky Way lies between the constellations of Cygnus and Scutum, and this is the most stunning part of the Milky Way. It is a collection of interstellar clouds – big clumps of dust and gas in between stars in the Milky Way – that provide the interesting colours you can see in many photographs of the galaxy from Earth.

There needs to be little to no moonlight to see the glow of the Milky Way branching across the sky. Remember, you need to let your eyes adjust for at least twenty minutes to see it in as much detail as possible.

The core of the Milky Way can only be seen around half of the year. It becomes visible around March and stops being visible at the end of November. The best time to see the Milky Way is the northern hemisphere's summer, or winter in the southern hemisphere.

Even in places where the light pollution is too strong, astrophotographers have been able to take photographs showing the beautiful glow at the centre of our galaxy.

Deep-sky objects

Most of the objects you can see in the night sky, especially with the naked eye and especially in urban areas where the skies are more polluted with bright lights, are stars. However, if you start to get interested in astronomy and knowing what lies beyond our planet, you will soon start to learn about deep-sky objects that are not stars.

There are three main kinds of deep-sky objects you can spot from Earth. Some of these can be seen with the naked eye, while others come to life through binoculars or telescopes. Broadly, these are nebulae, star clusters and galaxies.

Nebulae

Nebulae are clouds of dust and gas that have gathered together under gravity but are not dense enough for stars to begin to form yet. Eventually, the gas will collapse together, and stars will start to form. Until then, the gases stick together in nebulous clouds, hence the name. These are some of the most stunning things to look at in space, because the clouds form exquisite patterns and they are often photographed in different wavelengths, meaning the photographs of them are in incredible colours.

A nebula was first used to describe anything in space that was not closely packed together, and galaxies beyond our own used to be called nebulae. Now, it only refers to clouds of helium, hydrogen and other gases.

An interesting fact is that while nebulae are dense compared to the space around them, they are less dense than what we consider a vacuum on Earth. A nebula the size of our planet would weigh only a few kilograms.

The brightest nebula in our sky is the Orion Nebula, which can be found just below the belt of Orion. Through the naked eye, the nebula is visible, but its detail is revealed much more clearly with a pair of binoculars. The Orion Nebula has a magnitude of 4, which means even in skies with some light pollution it should be visible.

Other bright nebulae include the Lagoon and Omega Nebulae in the constellation of Sagittarius, and the Rosette Nebula. For other nebulae, you will need binoculars or a small telescope to spot them, but once you have a taste for gas clouds you will soon find your way to them.

These nebulae are all gaseous nebulae, but there are other types, too. However, gaseous nebulae are the only ones you will see with the naked eye. Planetary nebulae are shells of gas surrounding remnants of old stars, instead of gaseous nebulae which are related to stars being born. Planetary nebulae appear after a red giant star has shrunk and become a white dwarf. Their name was given because, through a telescope, they appear like a glowing disk which looks similar to a planet.

You might recall there are many ways stars can die. One of these – perhaps the most dramatic – is a supernova: the explosion and collapse of a giant star. After a supernova, the majority of a star's material is ejected into the space surrounding the leftover neutron star or black hole. This material looks similar to other nebulae but is created in a completely different way. The most famous example of a supernova remnant is the Crab Nebula, which has an apparent magnitude of 8.4, so you'll need binoculars to find it. It's in the constellation of Taurus.

Another type of nebula, dark nebulae, can have the effect of blocking out the light from stars beyond them, because the gas is so closely packed together. Most dark nebulae are found along the plane of the Milky Way, especially in the Great Rift, which we discussed earlier in this chapter.

Star clusters

Star clusters are exactly what they sound like: clusters of stars that are close together in space. There are two types of star cluster, globular and open. Globular clusters are so tightly packed that they are gravitationally bound to each other, while open clusters are less tightly packed and the stars do not affect each other's motion through their gravity. Globular clusters normally contain hundreds or thousands of stars, while open clusters tend to have fewer than 100. Open clusters tend to be easier to spot with the naked eye, because the stars are not packed so close together.

With the naked eye, we can see a few open star clusters from Earth – these include the Pleiades, Hyades and the Beehive Cluster. You might remember the Hyades from earlier in the book, in the section discussing the Taurus constellation (p. 164): the brightest stars in the Hyades cluster form the V-shaped asterism inside Taurus. The star Aldebaran is in the V-shaped asterism, too, but it is not part of the Hyades. The Hyades cluster is thought to be about 625 million years old. Its core stretches out almost nine light years across while the whole cluster reaches a span of thirty-three light years. Some other stars have also been spotted, further than this, escaping the gravity of the inner stars.

Another open cluster, also in the Taurus constellation, is the Pleiades, which is also called the Seven Sisters, because of its seven brightest stars. The Pleiades is one of the closest star clusters to Earth, making it visible with the naked eye and the easiest to spot out of all the star clusters. In Greek mythology, the Pleiades were the seven daughters of Atlas and Pleione, the sea nymph. Most of the stars in the cluster are hot, blue stars and astronomers think they are less than 100 million years old.

The Beehive Cluster, found in Cancer, is another close-by cluster, making it a great target for those with light-polluted skies. With a magnitude of 3.7, you won't see the Beehive Cluster in the middle of a busy street with lots of streetlamps, but get yourself into an open space like a park and the small, nebulous Beehive Cluster should become visible. To the naked eye, it looks like a single light source but, through binoculars, it becomes clear it is a cluster of stars.

Messier objects

The Hyades, Pleiades and Beehive Clusters, along with the Orion and Crab Nebulae, all have alternative names because they belong to a group of 110 different night-sky objects called Messier objects. Named after Charles Messier, all Messier objects have a Messier number, which is often given as M1 or Messier 1, for example. Charles Messier was obsessed with finding comets, and through his search for comets he came across a huge array of other objects that were really interesting, but not to him. In his anger at these non-comets, he wrote them down and categorised them.

Messier objects include star clusters, nebulae and galaxies, and they are a great target for any amateur astronomer. Some can be seen with the naked eye, while for others you need binoculars or a small telescope. Messier 1, or M1, is the Crab Nebula. The Orion Nebula is M42 and M43, the Whirlpool galaxy is M51 and the Andromeda Galaxy is M31.

Some astronomers embark on evenings called Messier marathons, in which they try to find as many Messier objects as they can in one evening. Some parts of the sky, like the Virgo Cluster and the Galactic Centre, are home to a large concentration of Messier objects, so they can be a great place to start. However, it is unlikely on any one night that all Messier objects will be visible at the same time. If you are interested, some websites list the best dates for the next ninety years in which to host a Messier marathon. Maybe something to aim towards.

Galaxies

Many of the Messier objects are galaxies beyond our
own. Galaxies are systems of stars clumped together
through gravity, and most have a supermassive black hole
at the centre which helps to stick them together, with its
immense gravitational attraction. There are three shapes
of galaxy – spiral, elliptical and irregular. The smallest
galaxies have just a few hundred million stars, while the
largest have one hundred trillion. Within the universe,
galaxies are organised together in groups, clusters and
superclusters.

In our galaxy, which is a spiral, there are about 100 billion
stars, and there are trillions of galaxies out there in our
observable universe. The galaxies closest to our own
make up a group called the Local Group, which contains
fifty-one galaxies. Beyond that, our Local Supercluster
contains about 100,000 galaxies.

We have only been certain of the existence of other
galaxies in the past 100 years, because it was in the
1920s that astronomers were able to distinguish between
galaxies and other nebulae. That means these trillions
of other stars out there were not known to us until a
century ago.

Floating around near the Milky Way there are small companion galaxies that orbit within our gravitational pull, called satellite galaxies. Two of these satellite galaxies, the Small Magellanic Cloud and the Large Magellanic Cloud are visible from the southern hemisphere with the naked eye. With a magnitude of 0.9, the LMC looks like a faint cloud with the naked eye, but it contains ten billion solar masses' worth of stars. It is found between the constellations of Dorado and Mensa. The smaller SMC is also visible from the southern hemisphere only, and even though it is fainter, with a magnitude of 2.7, is easy to see with the naked eye between the constellations Tucana and Hydrus. However, to get the best view it's advisable to look on a night with no moonlight, and away from light pollution.

Our closest large galaxy, which is not a satellite galaxy, is Andromeda or M31. As mentioned earlier in the book (p. 147), with its magnitude of 3.4, Andromeda is found near the constellations of Pegasus and Andromeda, and it's the only big galaxy that we can see with the naked eye. This makes it the most distant, and oldest, object you can see with your unaided eyes, at two and a half million light years away. Take a moment to think about that: when you look at Andromeda, you are seeing two and a half million years back in time. When the light that reaches your eyes left Andromeda, our ancestors *Homo habilis* were just appearing on Earth and the Ice Ages were beginning. Andromeda is much bigger than our own galaxy, with around one trillion stars, and it is the biggest in our Local Group.

In about four and a half billion years, it's thought our galaxy will collide with Andromeda, forming a huge elliptical galaxy. Until then, you can gaze up all you like and wonder about all that has happened in the universe between the light leaving those galaxies, stars or planets and when you are seeing them.

Other spiral galaxies include M66 in Leo and M83 in Hydra, which can be seen through binoculars since it's one of the brightest spiral galaxies. Elliptical galaxies are larger than spiral galaxies and have no spiral arms. Examples of these include M87 in the Virgo Cluster.

Even through binoculars and small telescopes, most galaxies just appear as faint patches of light, because of how far away they are. You need to look through a big telescope to see any detail of most of the galaxies out there. However, astrophotographers can use small telescopes attached to cameras to get amazing pictures of galaxies, but more about this in the Further Resources section at the end of the book.

Glossary

Apogee When something is orbiting something else on an elliptical orbit, the apogee is the point at which the distance between the two objects is smallest.

Apsis An extreme point in an elliptical orbit – either the closest or furthest point between two objects. Apogee and perigee are apsides (the plural of apsis).

Asterism A pattern of stars that is not an official constellation. Sometimes an asterism can be a part of a larger constellation, or it can be made up of stars from different constellations.

Asteroid A small rocky, metal body orbiting the sun, between Mars and Jupiter.

Astronomical Unit (AU) The distance between Earth and the sun (149 million km).

Aurora A display of light caused by particles from the sun interacting with gas in Earth's atmosphere, usually seen near the poles.

Azimuth The direction of an object in the night sky, measured clockwise from north.

Black hole A clump of mass so dense that gravity means nothing, even light, can escape.

Blue moon The second full moon in a calendar month.

Circumpolar A constellation that can be seen all year round: it never rises or sets.

Comet A rock made of ice and debris that orbits the sun in a long elliptical orbit, with the warmth from the sun vaporising its ice to create a tail.

Constellation A pattern of stars recognised by the International Astronomical Union, of which there are eighty-eight officially.

Corona The outermost layer on the sun's surface.

Coronal mass ejection A release of plasma from the corona.

Declination The equivalent of latitude on the celestial sphere, starting from north.

Eclipse A shadow cast from a planet or moon causing a second body to be obstructed from view. A lunar eclipse occurs when the Earth's shadow blocks the moon from view, and a solar eclipse occurs when the moon blocks the sun from being seen.

Ecliptic The path in the sky that the sun appears to follow, which is closely followed by the planets, too.

Electron A fundamental subatomic particle that carries a negative charge.

Equinox The time or date during which the day and night are of equal length. There are two equinoxes each year, in March and September.

Exoplanet A planet in any solar system other than our own.

Galaxy A collection of stars, gas and dust held together under gravity, often with a supermassive black hole at the centre.

Light pollution The glow of lights from near by that can obscure how many stars you can see.

Light year The distance light travels in one year, 6 trillion miles.

Magnitude The brightness of a star, planet or other object in the sky. Ranging from −25, the brightest, to 25, the dimmest. With the naked eye we can see objects of as low brightness as 5 or 6 in dark sky areas, but lower numbers are better.

Meridian The imaginary line that passes directly through north to south on the celestial sphere.

Meteor The flash of light seen when a meteoroid burns up in Earth's atmosphere.

Meteorite A piece of debris from an asteroid, comet or meteoroid that makes its way to the surface of Earth, or another planet or moon.

Meteoroid A piece of rock or metal travelling in space, usually small.

Meteor shower A spike in meteor activity that occurs when Earth passes through an area of space related to an asteroid or comet.

Milky Way Our galaxy, the Milky Way, is a spiral galaxy with about 100 billion stars.

Nebula Bright clouds of glowing gas that eventually will clump together to form stars.

Neutron star The star left behind when a huge star collapses after a supernova, a neutron star is around 30 km in diameter with a mass greater than twice our sun.

Opposition When a planet, moon or star is opposite the sun in the sky, making it visible during the night.

Parallax The amount an object appears to move as your perspective changes: for example, looking through your left eye compared to your right. It can be used to determine distances to stars.

Parsec A unit of distance used in astronomical scales, defined by the idea of parallax. A star one parsec away will have a parallax angle of one arc second compared to viewing it from the sun, one astronomical unit away.

Perigee When something is orbiting something else on an elliptical orbit, perigee is the point at which the distance between the two objects is greatest.

Photon A particle of light.

Planet An astronomical object orbiting a star that is big enough to be rounded due to gravity and have cleared its orbit of other debris, but not big enough for fusion to occur in its core.

Planisphere A chart that can be used to show what the sky will look like on a given date.

Plasma A fourth state of matter made of ionized gas and free electrons.

Precession The gradual movement of the Earth's axis of rotation, like a spinning top.

Proton A subatomic particle with a positive charge, made of three quarks.

Solar wind A stream of high-energy charged particles flowing out from the sun.

Solar flare A sudden flash of increased brightness on the sun near its surface.

Star An astronomical object made of plasma and held together through gravity, hot enough for fusion to occur.

Syzygy A straight-line formation of three or more astronomical objects: for example, the Earth, moon and sun.

Quark A fundamental particle, one of the building blocks of matter. Quarks cannot be broken down into anything smaller. They come in various flavours.

Zenith An imaginary point directly above any given point on Earth, on the celestial sphere. The zenith at the north pole is the north celestial pole.

Further resources

As with anything, advances in technology are making astronomy so easy for anyone who wants to learn about it. There are hundreds of apps and websites out there that provide great resources for people starting to get into astronomy, and it's best to try a few different ones out for yourself before committing to just one. The following lists are a suggestion of where to start.

Websites

Weather

One of the most important factors in stargazing is being able to predict the weather. The first thing you need to know is how much cloud cover there will be in your area. Secondly, the humidity and moisture in the air will tell you how 'twinkly' the stars are going to appear, and moisture can obstruct your vision. The following websites are a good place to start:

CLEAR OUTSIDE
https://clearoutside.com/forecast/50.7/-3.52

A weather website designed specifically for astronomers, Clear Outside gives you seven-day forecasts with hourly updates.

MET CHECK
https://www.metcheck.com/HOBBIES/astronomy.asp

Metcheck provides highly detailed cloud-cover forecasts for up to eight days ahead.

Dark-sky checkers

For anyone looking to stargaze in a city with light-pollution, the first thing you will need is a dark-sky checker. There are plenty of websites out there that will tell you how dark the skies are in your area and, importantly, how far you have to travel to get darker skies. The best way to check how many stars you can see is to go out and look, of course, but if you set your goal to be able to see more of the sky than you can in your city, these websites will tell you the best places to go.

DARK SITE FINDER
https://darksitefinder.com/maps/world.html#2/16.0/-302.2

This website has an interactive light-pollution map that spans the whole world. You can zoom in and see where the closest areas to you are with little light pollution – represented by the darker areas on the map. It can be handy before going on holiday, too, to see what the skies are like in the places where you're heading.

DARK SKY DISCOVERY
https://www.darkskydiscovery.org.uk/dark-sky-discovery-sites/map.html

Guides to the night sky

If you don't want to remember a long-winded star-hopping process and just want to use an app that shows you exactly how the night sky looks where you are, there are plenty of great websites out there for that purpose. The benefit of a lot of these websites is you can fast-forward to a particular date and time to see what the night sky will look like in the future, from anywhere in the world.

These websites can be a great place to plan your stargazing evenings, and to get to know the night sky. Of course, you won't be taking your computer stargazing with you, but many of these also have app versions if you want a reminder in your pocket.

STELLARIUM
https://stellarium.org/

This software is free to download on your computer, and it provides a complete planetarium – showing the sky just as it looks with the naked eye. There are also options to show what it would look like through binoculars or telescopes. Stellarium is available for Linux, Mac and Windows and it has a catalogue of over one million deep-sky objects, the constellations in twenty different cultures and more than 117 million stars. What more could you ask for?

Meteor showers and solar eclipses

The following websites are a few that will give you great guides to the meteor showers coming up over the next few months. They will also provide information about where to see total and partial solar eclipses.

EarthSky.org

TimeAndDate.com

StarDate.org

Space.com

Apps

Sometimes you will be caught looking at the sky, wondering what it is you can see. On nights like this, it can be handy to have astronomy apps on your phone to help you identify what you are seeing even when you aren't dedicating a whole evening to stargazing. This is especially true if you happen to catch a planet looking particularly bright one night, or a meteor shower.

Most of the websites mentioned previously have dedicated apps too, so have a look to see if you can find your favourite. Otherwise, explore the following list for plenty of resources.

Weather

SCOPE NIGHTS
https://eggmoonstudio.com/

This app takes weather data based on your location, to which you can either allow it access or you can input a location and it will interpret it to tell you how good the conditions will be for stargazing. It also has a dark-sky map.

Night-sky guides

All of these apps let you point your phone in front of you

and immediately identify the stars, satellites and planets you are looking at. Most have a free version and you can upgrade to a more premium, paid version without adverts if you like. Try out a few and see which interfaces you prefer.

SkyView

Terminal Eleven

Star Walk 2

Star Rover

Star Chart

Night Sky

SkySafari

Mobile Observatory

Pocket Universe

Sky Map

Northern/Southern lights

As mentioned in the Aurorae section, there are some apps that help you use your phone to identify the aurorae before your eyes spot them. They can also give you alerts and weather forecasts for your area, to increase your chances of being able to see them. If your phone camera allows you to adjust the settings, you want your exposure to be high and a long shutter speed, to be able to see the green glow.

FOR PHOTOS OF THE NIGHT SKY IN GENERAL:
NightCap Camera

Amateur Astrophotography

FOR WEATHER AND AURORA ALERTS:
My Aurora Forecast

Aurora Alerts

Aurora Notifier

WEBSITES:
NOAA/ Space Weather Prediction Center

N3KL.org

SpaceWeatherLive.com

More information

NASA
This app is not for stargazers, but it is packed full of information about space, including the latest news, videos and updates on missions.

SPOT THE STATION
For those keen to see the space station, this app is dedicated to helping you. It will give you alerts whenever the ISS is visible from where you are. It's always visible in cities, too, so urban astronomers should take note.

SOLAR WALK
With this app, you can take a journey through the solar system in 3D, learning about the planets as you do so.

MOON ATLAS

This app lets you explore the moon with your fingertips. A detailed picture of the surface of the moon, you can zoom in and see all its craters.

SKYWIKI

This app has a great user interface and is full of information for stargazing. It gives lots of information about astronomical events, both past and future. There is also a page that shows an Astronomy Picture of the Day, NASA image of the day, lunar image of the day, and Hubble picture of the week.

SPACE IMAGES

Pretty much what it says on the tin, this app will show you a new photo of an amazing space phenomenon every day.

Other resources

Astronomy clubs

If you are interested in astronomy after reading this book, it's a great idea to join a local astronomy club. There, you will meet like-minded people who will be able to teach you the best ways to see the night sky in your city. A quick search online will let you see whether there is a club near you, and what activities they do. The club will also have an idea where the best spot for stargazing in your city is. For example, in London there is a popular astronomy club that meets in Regent's Park. Astronomy clubs can be daunting, so make sure you find the right group for you and take a friend along if you don't want to be overwhelmed. It'll be OK, though – astronomers are friendly people.

Museums, observatories and planetariums

Don't forget to check out the places that are dedicated to communicating science in your area. Museums, especially science museums or science centres, can be a great place to learn more about space and stargazing. Local observatories and planetariums often run events where people can go along and learn about the night sky, with some time looking through their telescopes, too. It can be another great activity to add to a holiday, for a slightly different evening, so when you are travelling make sure to research whether there are any observatories near the places you plan to go.

Magazines

Magazines are another great way to keep up with what is going on in astronomy. From telescope recommendations to exciting events and science news, astronomy magazines are a great resource. Many also carry offers for money off equipment, so if you think you're after a telescope, for example, it could be worth subscribing to one. Here are a few of the most popular:

Sky and Telescope

Astronomy

Astronomy Now

All About Space

BBC Sky at Night

Astrobiology

SkyNews

SpaceWatch

Astronomy Technology Today

Go Taikonauts Magazine (for those in China)

Astrophotography

One of the most enticing aspects of astronomy is astrophotography, because – let's face it – these days, does it even count if you didn't get a photo of it? As the first chapter of the book suggested, though, it is important to get to know the night sky a little before you invest in any equipment for photographing it.

Thanks to improvements in technology, anyone can pick up a decent camera and tripod and start taking photographs of the sky. This book is not about astrophotography, and if you want to get started with it the best thing is to join an astronomy club and meet people who are already taking great pictures of the night sky. For an example of what can be done, have a look at the Royal Astronomical Society's competition, Astronomy Photographer of the Year. Not only will the qualifying photos blow your mind with their beauty, they are a great example of what can be achieved with different kinds of set-ups. Each photograph is presented with details of how it was taken, what kind of camera and shutter speed were used, etc., so you can see exactly what is achievable.

The benefit of astrophotography in cities is that there is a lot of equipment you can buy that helps block out the light pollution, so the camera can see a lot more than you can. There are a few apps that mean you can take photographs of the night sky even on your phone, but for the best photographs you will at least need a tripod to hold the phone still.

Once you are into advanced astrophotography territory, you will want to buy a small telescope to attach to your camera, along with a motorised tracker so that you can take long exposures but keep the camera focused on one object. Essentially, the trackers mimic the rotation of the Earth but in the opposite direction, so your telescope stays fixed on whatever point you start it on. These set-ups are not cheap, so you should be sure you want to take lots of photographs before you invest.

Sources and references

Introduction

https://www.un.org/development/desa/en/news/
population/2018-revision-of-world-urbanization-
prospects.html

Uzan, J-P; Leclercq, B., *The Natural Laws of the Universe:
Understanding Fundamental Constants.* (Springer,
2008), pp. 43–4

Aughton, Peter, *The Story of Astronomy* (Quercus, 2011)

https://www.space.com/17738-exoplanets.html

1. Everything except stars (the closer stuff)

https://www.northernlightscentre.ca/northernlights.html

https://www.nasa.gov/content/about-auroras

https://www.swpc.noaa.gov/products/aurora-30-
minute-forecast

https://www.jpl.nasa.gov/infographics/infographic.view.
php?id=11268

https://www.webcitation.org/6ftO4K7lC?url=http://
nssdc.gsfc.nasa.gov/planetary/factsheet/venusfact.html

https://solarsystem.nasa.gov/news/813/nasas-cassini-
data-show-saturns-rings-relatively-new/

https://solarsystem.nasa.gov/planets/uranus/overview/

Stern, A., and Mitton, J., *Pluto and Charon: Ice Worlds on the Ragged Edge of the Solar System* (Wiley-VCH, 1997, second edition 2005)

Jenniskens, P., *Meteor Showers and Their Parent Comets* (Cambridge University Press, 2006)

http://coolcosmos.ipac.caltech.edu/ask/150-What-is-the-weather-like-on-Neptune

https://image.gsfc.nasa.gov/poetry/venus/q89.html

https://www.nationalgeographic.com/science/space/solar-system/full-moon/

https://www.space.com/35526-solar-system-formation.html

https://space-facts.com/dwarf-planets/

https://www.webcitation.org/6ftO4K7lC?url=http://nssdc.gsfc.nasa.gov/planetary/factsheet/venusfact.html

https://sites.google.com/carnegiescience.edu/sheppard/moons

https://www.wired.co.uk/article/nasa-reveal-discoveries-cassini

http://blogs.discovermagazine.com/d-brief/2019/01/17/age-formation-saturns-rings-cassini/

https://space-facts.com/uranus/

https://www.nasa.gov/audience/foreducators/9-12/features/F_How_Far_How_Faint.html

https://www.geek.com/feature/11-awesome-facts-about-the-lovable-dwarf-planet-pluto-1629287/

2. How to stargaze

Thompson, M., *Philip's Stargazing with Mark Thompson: The Essential Guide to Astronomy* (Philip's, UK ed. Edition, 2015)

Dunlop, S. and Tirion, W., *Collins Night Sky & Starfinder* (Collins, 2011)

https://www.nationalgeographic.com/travel/lists/activities/tips-for-better-stargazing-astronomy-viewing/

http://www.astronomy.com/observing/get-to-know-the-night-sky/2006/12/setup-is-key-when-urban-stargazing

http://www.nhm.ac.uk/discover/how-to-stargaze-in-cities.html

https://pineriverobservatory.wordpress.com/suggestions-for-entering-amateur-astronomy/urban-stargazing-whats-possible/

https://www.timeanddate.com/astronomy/seasons-causes.html

https://imagine.gsfc.nasa.gov/features/cosmic/milkyway_info.html

http://abyss.uoregon.edu/~js/glossary/stellar_magnitude.html

https://lco.global/spacebook/what-apparent-magnitude/

https://www.spaceanswers.com/deep-space/supergiant-stars/

https://www.skyandtelescope.com/astronomy-news/spinning-pulsar-smashes-record/?c=y

https://www.iau.org/public/themes/naming_stars/

https://www.space.com/15567-north-star-polaris.html

http://hubblesite.org/image/1842/news_release/2006-02

https://earthsky.org/brightest-stars/star-errai-future-north-star

https://www.constellation-guide.com/the-southern-cross/

http://www.wetheitalians.com/default/great-italians-past-amerigo-vespucci

https://books.google.co.uk/booksid=LvnNFyPAQyUC&pg=PA94&lpg=PA94&redir_esc=y&hl=en

3. What to see – the stars

https://earthsky.org/astronomy-essentials/

http://www.ianridpath.com/starnames.htm

https://www.iau.org/public/themes/constellations/#cnc

https://www.iau.org/public/themes/naming_stars/

https://www.newscientist.com/article/2107207-no-nasa-hasnt-changed-the-zodiac-signs-or-added-a-new-one/

Bakich, Michael E., *The Cambridge Guide to the Constellations* (Cambridge University Press, 1995)

Dunlop, S. and Tirion, W., *Collins Night Sky & Starfinder* (Collins, 2011)

Evans, J. *The History and Practice of Ancient Astronomy* (Oxford University Press, 1998)

Ridpath, I and Tirion, W., *Stars and Planets Guide* (Princeton University Press, 2001)

Thompson, M. Philip's *Stargazing with Mark Thompson: The Essential Guide to Astronomy* (Philip's, UK ed. Edition, 2015)

4. The Milky Way

Levy, D. H., *Deep Sky Objects* (Prometheus Books, 2005)

https://web.archive.org/web/20110722021459/http://blackholes.stardate.org/directory/factsheet.php?p=M31

http://www.backyard-astro.com/focusonarchive/m31/m31.html#Anchor-11481

https://www.eso.org/public/images/potw1706a/

https://www.wired.co.uk/article/what-happens-when-two-neutron-stars-collide

Credits

Trapeze would like to thank everyone at Orion who worked on the publication of *The Art of Urban Astronomy* in the UK.

Editorial
Grace Paul
Jennifer Kerslake

Copy-editor
Sarah Hulbert

Proof-reader
Patrick McConnell

Audio
Paul Stark

Contracts
Paul Bulos

Design
Rachael Lancaster
Helen Ewing
Briony Hartley

Finance
Jasdip Nandra
Elizabeth Beaumont
Sue Baker

Marketing
Tom Noble

Production
Claire Keep
Katie Horrocks

Publicity
Leanne Oliver

Sales
Jennifer Wilson
Esther Waters
Deborah Deyong
Victoria Laws

Rights
Susan Howe
Richard King
Krystyna Kujawinska
Jessica Purdue
Hannah Stokes

Operations
Jo Jacobs
Sharon Willis
Lisa Pryde

Acknowledgements

Thank you to my lovely editor Grace Paul for working with me on this book, and to everyone else at the Orion Publishing Group who has helped put it together, especially Jennifer Kerslake, Sarah Hulbert, Patrick McConnell and Helen Ewing.

To the endless list of physics and astronomy researchers who have answered my questions over the years, thank you. It's not your job to speak to journalists but you do, and it's appreciated.

This book wouldn't be here without the support from my amazing family and friends. Thank you to my mum, for your unwavering support, and to my dad for always being there to discuss physics and the stars with me. Thanks to James for introducing me to Muse all those years ago and watching Supermassive Black Hole live with me while I lost my shoe aged fifteen. Thank you to Grandma Beall, for everything, my lovely aunties Anne and Jane, and my cousins Clare, Vicky, Natalie and Laura. To Darcey, Lola, Max and Maia, I hope you will grow up and enjoy this book for a long time. To Gus, Kris, Robb, Janice and the rest of the Mathies, I love you all, and this book is for Steff.

Thank you to Rachael for being there always even though you're on the other side of the world, and for believing in my ability to write a book all those years ago. Thank you to Vicki, you have been there through the best and worst of times with me, and to James and Abby for getting me through our years of studying physics. Not sure I'd have a degree without you guys. To Anna and Dan, thank you both for being a true support in my life and my work. Thanks also to Scott, Paul, Jessica, Carolyn, and Phil – you put up with my space chat for so long and are still there to stare at the sky in awe with me (in between ceilidh dances, of course).

Finally, the most thanks go to my favourite stargazing and travel partner, Joel. Thank you for sharing moments I will never forget – from spotting our first galaxy, seeing the moons of Saturn through a telescope in New Zealand, to cloudy and rainy nights in the north of England. Hopefully there will be many more adventures to come.